CW01662414

Conten

Printed by
in Great Britain

Editor
Greg Payne

Published by
Greenlight Publishing
The Publishing House, 119 Newland Street,
Witham, Essex CM8 1WF
Tel: 01376 521900 Fax: 01376 521901
email: info@greenlightpublishing.co.uk
www.greenlightpublishing.co.uk

ISBN 1 897738 110

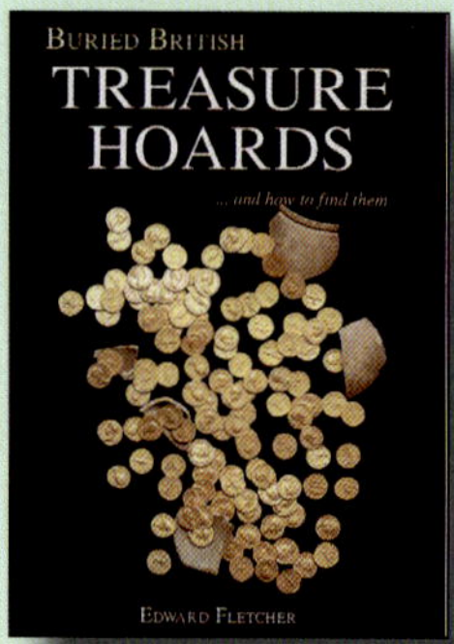

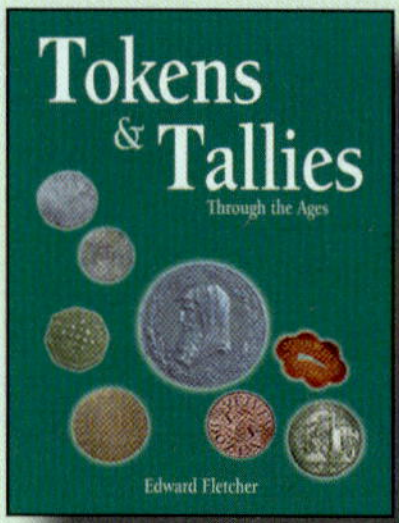

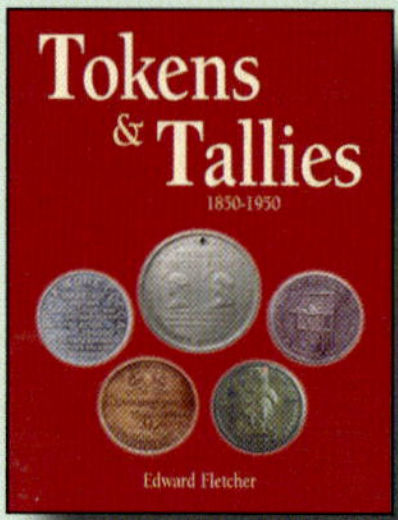

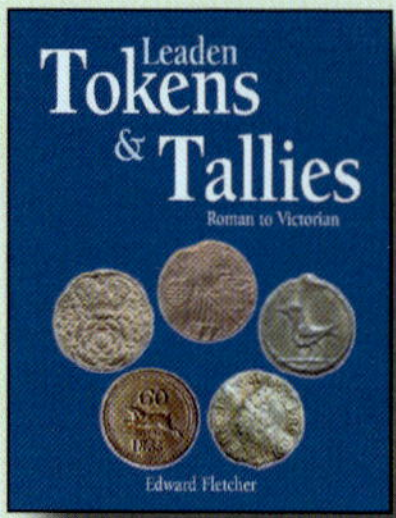

Also available from Greenlight Publishing
by the same author
www.greenlightpublishing.co.uk

☎ **01376 521900**

Introduction

"Knowest thou the way to Dover?"
"Every stile, every gate, every drove, every horseway, every footpath."
(King Lear).

William Shakespeare penned those lines around AD 1600, in an age when poor to middling folk went about their daily business on foot, while richer merchants and aristocrats felt proud to own a horse. The swiftest means of getting goods and commodities to distant markets in those days was across the back of a mule along what we now call bridleways; and anyone selling flesh or fowl sent it live and on foot via a network of drovers roads that criss-crossed the land like veins and arteries in a human body. Little had changed in the previous 1600-odd years of everyday British life, save for language, which had evolved from Celtic, via Latin and Anglo-Saxon, to the near-modern version of English in which the Bard wrote his immortal works.

Let me remind you that during those many centuries nobody owned an A-Z street map or an AA road atlas. Instead, every man, woman and child possessed an encyclopaedic knowledge of the stiles, gates, droves, horseways and footpaths in their neighbourhood. Drove masters and mule train captains knew even more. They could safely transport several hundred black cattle from the Highlands of Scotland or the mountains of Wales to the fattening grounds of East Anglia and Kent, or even to Smithfield at London's heart. And they could confidently carry bales of cloth and wool from northern regions to the emporiums of city merchants. Their knowledge of routes, watering places, crossroads, safe overnight encamping grounds, fords, welcoming inns and prominent landscape features in every county must have been phenomenal.

Modern life has rendered such knowledge largely obsolescent, but as detectorists we begin to appreciate its importance when the golden rule for success at finding coin and artefact losses and throwaways dawns not long after we buy our first detector: the majority of finds are made at places where people gathered in the past, and the greater the number of people at any place, whether in a large crowd infrequently, or in smaller numbers but very often, the greater the quantity of coins and artefacts lost or discarded at that place.

I now refer to these locations as "nodal points of communication". The simplest example I can think of would be a large field crossed by three footpaths, all converging on a stile in a corner of the field. The nodal point is obviously the stile; all the people who ever crossed the field climbed the stile; some met at the stile; small groups might even have discussed farming/village gossip at the stile; local drunks might have stumbled at the stile on their way home from a cider-supping session; lovers probably fumbled there on warm spring evenings. Over time the numbers must have run to hundreds, perhaps thousands of people converging on the stile. A detector swept over the ground up to several yards around the stile is almost bound to detect losses.

A less obvious nodal point is the outer edge of the perimeter wall surrounding any old churchyard. To appreciate its "nodality" you must be aware that in times past members of the congregation would almost certainly have gathered at the wall after church services. Victorian manners restricted this custom to shaking hands with, or curtseying to, the vicar; but in pre-Victorian times gossiping, courting, playing games, even holding markets were commonplace activities around churchyard walls. I've witnessed dozens-upon-dozens of hammered coinage finds along the bases of churchyard walls; ironically, the sovereign whose head has featured most often on the coins I've seen brought to light in this way has been Henry VIII, the monarch responsible for so much ecclesiastical destruction during his reign.

In this book I have listed and/or illustrated about 50 assorted nodal points and explained why they provide such good clues when tracking down spots where coins and artefacts lie thickest in the ground. You can locate your own nodal points, perhaps with slight regional variations, in every county of Britain. In addition to a copy of **Reading Land** you will need 1:25,000 Explorer Ordnance Survey maps for your chosen search area. If you take those maps to a local reference library or county archive you can compare them to older, perhaps Victorian editions of large-scale OS maps and make pencilled notes on your own maps whenever you see potentially interesting features, structures or find spots recorded on the earlier sheets. In some cases the notes you make will probably put you on the track of a nodal point; in other cases you might suddenly appreciate the significance of a particular feature as a nodal point when studying your marked maps. Later you can apply that knowledge to identify similar sites in other districts.

Binoculars, a compass, a camera and lightweight walking gear all feature in my kit when I'm on a trip seeking historical clues in the countryside. The binoculars come in useful for everything from scanning the surfaces of ploughed fields for pottery fragments, to looking for ancient building materials re-used in church spires. They also make it possible for me to survey quite large tracts of ground from one convenient vantage point. My compass does more than indicate north and keep me from getting lost. It also shows me south-facing slopes of prime agricultural land. Such places are nodal points because down many centuries they have attracted colonists and farmers, thanks to their sunny aspects, warmer soils and excellent crop-growing potential. I carry a small camera because there's nothing better than a photograph of one site when it comes to winning permission to search another site.

On the subject of permission, I have to point out here that, unlike the tidal Thames where one can obtain search permission for the payment of a small licence fee, the land you hope to detect on will almost always have a private owner from whom you must obtain permission in advance of your search. Take photographs beforehand from a public road or footpath if you feel that showing photographs to the landowner will help your cause when it comes to gaining permission to search. Be ready to explain your interest in local history and in locating nodal points. You will have become an expert on the subject by the time you reach the end of this book, though let me add that this book by no means exhausts the subject.

I would be happy to learn of your successes on or close to other nodal points not included in the pages that follow.

Please contact me via E-mail: fletchnews@aol.com.

As with **Reading Tidal Rivers**, this book is aimed primarily at relative newcomers to the hobby, with a nod of recognition to old hands at the game who will, I hope, find one or two snippets worth adding to their already extensive knowledge of site finding. Even dedicated seekers after Celtic, Roman and Anglo-Saxon coins and artefacts might experience a few pleasant surprises when ancient finds turn up on some of the sites discussed as potentially productive of medieval to modern losses. That is, of course, because fertile soils and locations favoured with good sunlight, drainage, water and micro-climate have always attracted humans who need to make a living from agricultural land; Celtic field makers, Roman villa builders, Anglo-Saxon settlers, Norman landlords - all took some interest in the soil now occupied by many of the most prosperous modern farms.

Here then, is my selection of nodal points of communication where I predict that relative newcomers to the hobby, as well as one or two old hands, will make interesting finds in their immediate vicinities. They are not set out in any order of importance, but if you can locate a combination of two or more nodal points occurring close together you will have even better prospects of success. Good hunting!

Fig.1(a)

1. Barns

Anglo-Saxons called them *bere-arns* (=barley houses) and needed them because developments in post-Roman agriculture yielded crop surpluses that required storage under cover. Norman lords built them to hold excess production achieved with a labour force of serfs. The medieval Church authorities required one in every parish as a collection centre for tithes due to the local priest.

From Tudor times down to Victoria's reign barns became increasingly multi-purpose buildings used as much for village social events as for farming tasks - barn dancing, for example, was popular in English villages more than a century before Oklahomans took it up in America. Barn doors were always built as wide as possible and in pairs, on opposite sides of the barn, and strategically placed to catch a strong draught when both sets of doors were open. The force of the wind helped to dry any root crops stored on the barn floor between the doors. Similarly, the draught was an aid to winnowing grain; chaff blew out of the barn in a downwind direction and was eagerly pecked up by hens and pigeons flocking around the door in anticipation of the feast.

All of these activities made every barn a communication node with many hundreds of people working and playing in and around the building every year.

Most of the footpaths leading to an old farm arrived at one of the barn doors. The footpaths will be preserved on modern maps, so even if you can't find an ancient barn in as good a state of preservation as that shown in Fig.1a., you

Fig.1(b)

should be able to locate the sites of many by reference to old large-scale OS maps. On an early 6in map the largest building on any farm will probably be its barn. Make a larger scale sketch of the buildings and show the sketch to the present farmer. He will probably allow you to try your luck at detecting on the site of the ancient barn if the ground remains accessible. Barns had earth floors so the ground will be rich in casual losses.

Fig.1b. shows part of the internal fabric of a very old barn. All timber used in its construction would have been locally grown, cut and sawn. In northern and western parts of Britain barns were more often built with stone walls, though roofing timbers would still have been locally grown. Old stone barns are more likely to survive the passage of time, even if used as little more than large garages by some present-day farmers. If the floor is not concreted the site could still yield good finds.

Browse through the Historic Monuments records for your county. Visit the sites of any listed barns, if only to view them so that you become acquainted with such structures. Even if you can't detect within the barn you should still try winning detecting permission on the adjoining farmland. At some time in the past it must have been an important farm; this will reflect in what lies in the local plough soil.

2. Cropmarks

Because cropmarks don't photograph particularly well at ground level, I had to select some very obvious cropmarks for Fig.2. which depicts the visible traces of a Victorian farm labourer's cottage abandoned "before the First World War" according to a local octogenarian farmer.

Nevertheless, the photograph is useful because it illustrates some important aspects of cropmark hunting, including crop colour, crop growth, bare patches, weed growth, even misleading marks in growing crops. This field has experienced intensive ploughing for three quarters of a century since Edwardian times, but the location of the old building is plainly visible at most seasons of the year when any kind of crops grow on the land. The brick walls of the cottage offered a useful source of free materials for pig sties built on the farm in the 1920s; broken bricks and other rubble became hardcore beneath concrete yards near the main farmhouse; doors, roof beams and window frames were recycled; other wood was chopped for fire-lighting. And in the 1970s all hedges around the cottage garden vanished (together with the hedgerows from six nearby small fields) when the farm became a mono-crop prairie growing almost nothing but wheat. It has since reverted to mixed farming and this field had clover growing in it at the time I took the photograph.

Look first at the parallel lines running right and left out of the picture. These are tramlines - the ruts into which a tractor's wheels, and the wheels of any machine towed by the tractor, fit neatly. The same tram lines will be used when sowing, applying weedkillers or fertilisers, even when laying out irrigation pipes if the crop needs watering. Don't allow tram lines to deceive you into thinking you've found an ancient track or road, which leaves a quite different crop mark signature, described below.

Look next at the weeds standing higher than the main crop in the middle distance. Despite applications of weedkiller, these tough specimens survive because they have sent strong roots into the deeper soil usually found where old hedgerows once grew. You can trace the outlines of old fields thanks to lines of weeds. Mark them carefully on your Explorer map and you will have potentially

Fig.2

productive areas to work when the field is ploughed and harrowed; farm workers and local villagers would have sat along the old hedges in centuries past.

Parallel rows of weeds (not tram lines) might indicate ancient lanes, trackways and droves. If these converge on a particular location detect that area carefully; converging routes always brought people together at a nodal point.

Note the darker crop growth in the central area of the photograph. Disregarding the tramlines, the deeper green can be seen as a fairly clear oblong. It marks an area once occupied by the cottage garden, deeper green thanks to rich nutrients and additional moisture in the ground hereabouts. This area will also be much richer in losses because the garden provided a nodal point that drew the farm labourer, his wife and children - all working or playing in the back garden of the cottage. One could confidently expect finds dating back to at least Georgian times on such a patch after ploughing and harrowing.

Next, look at the bare ground and the area where hardly any crop growth has occurred. It extended a few yards beneath my feet from where I stood to take the photograph. Oblong in shape, it accurately matched the original foundations of the cottage. I tried thrusting a trowel into the ground here, but it proved impossible because foundation rubble exists just a fraction below the surface. Two thousand years into the future the combined effects of earthworms, ploughing and wind erosion will have deposited a couple of feet of topsoil on the foundations. They will be buried by soil that comes from adjacent ground. I hope that point makes my message clear: detecting on or over buried foundations is unproductive. Concentrate on hedgerow lines and on areas where more lush crop growth indicates former deep cultivations.

Seekers after Celtic, Roman and Anglo-Saxon losses often pay close attention to cropmarks because they certainly do indicate numerous ancient sites. But detectable finds will always be greater in number in the fields now flanking ancient structures for precisely the reasons given above. Structural remains will probably have protected status as ancient monuments; but don't allow that fact to deter you from seeking permission for searches on flanking fields. Beginners bent on ancient coin and relic hunting should find a copy of **Treasure Hunting** dated May, 1996, in which contributor Chris Delamont gave a fascinating account of cropmark hunting with a camera and a model aircraft. Equally good results might be obtained by anyone who can gain access to church towers or hilltops from which fields in crop can be scanned with binoculars.

Fig.3(a)

3. Reused Building Materials

Northumberland offers one of the most striking examples of early reuse of building materials. Literally hundreds of old farmhouses, dozens of Anglo-Saxon and Norman churches and mile-upon-mile of medieval boundary lines on both sides of Hadrian's Wall have stonework robbed from the famous Roman monument. Inscribed tombstones, barrack-block paving slabs, even carved latrine seats have all been discovered serving quite different purposes in later structures. Who can blame those who saw an old, tumbledown wall apparently serving no useful purpose as a source of ready-made building materials? And why should we not expect the same thing to happen in other parts of the country to abandoned Roman

Fig.3(b)

villas, or to the buildings in a deserted medieval village, or any Georgian packhorse bridge superseded by a Victorian railway arch?

Look carefully at the walls of any obviously old buildings on farms and in villages in your area. You may spot Roman or medieval bricks reused, as in Fig.3a. Seek out the owner/occupier and ask if they know where the reddish bricks or tiles in the walls came from. Is a local field noted for having particularly stony patches? Or have chance finds of ancient date turned up on a nearby plot of land? Even if your questions result in negatives, don't forget to look for other clues in the neighbourhood - converging footpaths, south-facing slopes, etc. Those ancient bricks must have come from a local source.

Fig.3(c)

Fig.3b. shows ancient stone reused in a similar way. Gate posts, milestones and door lintels often turn out to have been worked by stone masons long dead before the present structure went up.

Fig.3c. shows stones gathered from the fields of one small farm and reused with a railway sleeper to make an attractive seat. The stones hint at medieval to Victorian structures on the land in the past.

Fig.4(a)

4. Pollards & Coppices

It has become increasingly difficult to find living examples of pollards and coppices simply because woodland crafts and skills have largely died out; but the little oak tree in Fig.4a. shows what was once a common sight all over Britain from Anglo-Saxon times until the exploitation of coal on a large scale in the early 19th century - a stumpy hardwood tree with vigorous new growth sprouting from its top.

The trunk of the tree is about 8ft in height; just too long for grazing cattle and deer to get at the leaves. When lopped or "polled" in this way a tree would bring forth numerous

Fig.4(b)

shoots to start new branches. In a couple of seasons the young growth could easily be harvested to provide thin hardwood ideal for charcoal burning or for building purposes. Charcoal made in this way had a variety of uses from fuel, to gunpowder, to the by-production of pitch for caulking the timbers of barges, boats and sailing ships.

Any pollarded tree would yield a crop every few years, and it would go on doing so for centuries. Examine early 6in Ordnance Survey maps and you'll spot hedgerow pollards in every county. They often became boundary markers, mentioned in legal documents and recorded on ancient estate plans. People worked on or near them for hundreds of years, so fields flanking lines of old pollards, or the now vanished hedgerows where pollards once stood, make excellent detecting sites.

A coppiced tree or shrub (beech, hazel, hawthorn, etc) was cut back to ground level each autumn. In spring it would send out new growth and provide a harvest of long, thin twigs for basket and hurdle making. Fig.4b. shows the remnants of a coppice now overgrown and neglected. A century ago entire woods were coppiced. The craft called for teams of workers to cut and harvest the crop. Coppices were marked as such on early Ordnance Survey maps, so it's not difficult to identify nearby field boundaries and footpaths leading to the site of a former coppice. Those are the spots on which to concentrate; that's where coppice workers would have rested under hedgerows and lost possessions.

Fig.5(a)

5. Vanished Industries

"This village was once a centre for the ..... industry." Time and again you will read those words as you browse through a book on your county's history. Depending on which part of the country you live in, the vanished industry might be lead mining, hop roasting, weaving, dyeing, leather tanning, pottery manufacture, gravel digging or any one of dozens more. Each needed raw materials, labour, customers and communications via lanes, roads and/or rivers.

Figs.5a, 5b. & 5c. show clues to some vanished industries spotted during a few hours spent cycling around villages in one county. Carry out your own survey, on the lookout for buildings that have converted from some form of industrial use to modern homes or offices. Mark their locations on your OS maps and try to figure out from footpath and road alignments how the products of the industry were probably transported. Link each site with a main road, or with a local riverside.

The best of the bunch will have arable fields right alongside. Concentrate your searches on such fields and you will soon find relics of the former industry - tallies, tokens,

Fig.5(b)

tools, money and personal possessions lost by workers decades or centuries ago.

If luck is on your side, a local history book might record all you need to know about the vanished industry, but with or without a book you should always introduce yourself to the present occupants of the building. They may also own adjoining land; interest in their home's history is sure to bring a search-and-share agreement for you and your detector.

Fig.5(c)

Fig.5c. shows old gravel workings abandoned in the late 19th century. Arable fields adjoining sites like this always deserve attention, if only because gravel soils are light and easy to work and usually well drained - factors that made them attractive to early farmers. But even without evidence of ancient habitation nearby, footpaths leading to old gravel workings can still repay attention. Additionally, there is always a possibility that old pits such as these were used in Victorian or Edwardian times as refuse dumps. A partially filled-in pit could hold "glass gold" in the shape of antique bottles. Old workings have been put to many uses; I know of a site in Essex that began life as a gravel pit, but then became a grassy amphitheatre where local pageants took place. Today it serves as a local recreation ground.

6. Windmills & Watermills

Thanks to the Black Death and a massive shortage of labour that followed its passage across Britain in 1348, most millers managed to free themselves from servile conditions imposed by Norman landlords in earlier centuries. By 1400 they were largely independent, though probably still paying ground rent and taxes to the lord of the manor. Their mills, always nodal points of communication, now witnessed frequent cash transactions, with inevitable losses into the ground in their vicinities.

The number of working mills in a medieval parish depended on the fertility of local soils: lots of corn meant several mills. Whether these were windmills or water mills was also determined by natural factors. A windmill could grind corn more efficiently, but it needed an exposed site and a predicable wind. The main requirement for a watermill was a reliable and sufficiently powerful flow of water in an accessible river or stream.

The **Doomsday Book** will tell you how many watermills your chosen parish had, but windmills arrived in the 14th century, too late for mention in that work. Nevertheless, you'll probably find references to any windmills in a well-researched local history book. The distribution of early footpaths provides further clues to mill sites. If two or more paths converge at a river or streamside, but you can see no corresponding paths on the opposite bank, mark the area as a probable watermill site. Even though numerous watermills were either swept away by floods or destroyed during later river improvement work, the telltale footpaths leading to the mill remain as clues. A nearby pond might be the former millpond; tributary streams may once have formed the millrace.

Detect all available surrounding land and don't be surprised if you make finds of Roman date because medieval watermills often reused the sites of Roman dams, taking advantage of available Roman stonework. Keep a lookout for Roman pottery, bricks and tiles on nearby fields.

More than 7,000 watermills were recorded in the **Doomsday Book**. They were all employed in grinding corn, but in later centuries the power of the wheel was also harnessed in other industries - wool, cotton, snuff, hops,

Fig.6(a)

and more. Adjacent land will hold relics of those days and trades.

By 1800 the number of working watermills in Britain had risen to 40,000. Steam and electricity began to supersede them after 1850. Anticipate finds of Roman to Victorian date close to any former watermill site.

There's a wide misconception that windmills were always sited on hilltops. Put yourself in the boots of a medieval farmer with a cartload of corn and a need for flour. The last thing you and your overworked oxen want to face at the end of harvest is a strenuous climb up a steep hill. No, what the miller looked for when planning a new mill was an exposed site that could be reached over gently sloping

Fig.6(b)

ground. That's where you should seek windmill sites. Study the contours on your Explorer map; look for even spacings between rising contours. A more obvious clue might be the name Windmill Hill or Windmill Lane. Names such as Castle Hill and Barrow Hill have also been given to windmill sites because windmills were often built on artificial mounds to gain extra height for the sails. Such mounds survive long after the timbers of a mill have vanished, leaving what can look like an ancient burial mound or the site of a Norman motte. A derelict or reused windmill of Victorian vintage gives another good clue to the location of medieval windmills. If the site was suitable for a Victorian mill it probably attracted earlier millers. In some favoured

Fig.6(c)

spots the remains of half a dozen vanished windmills spanning centuries of corn grinding have been identified by mounds of earth or piles of stones. They acted as nodal points throughout their working lives.

Almost all Victorian windmills that survive today as intact structures now serve as homes or offices (as in Fig.6a.); or as museums (as in Fig.6b.). If you can gain access to any open ground around them you could still make good finds. Fig.6c. gives some impression of the stonework used in making a millrace for a watermill. The larger stones might have been salvaged from a former Roman dam. When you can find no visible remains of a watermill on a suspected site, check local buildings for reused stonework that might confirm a mill's existence.

Fig.7(a)

7. Crop Recognition

The majority of readers will recognize the barley ears in Fig.7a., but how many can name the crops shown in Figs.7b., 7c., 7d. and 7e.? Do you also know the sowing, growing and harvesting routines for each crop? What about soil preparations before the seeds go into the ground? And which crop is likely to follow this one; how soon after harvesting will the farmer begin to prepare the ground for the next crop? You must have answers to all of these questions. Why? Because your painstaking work on researching and identifying nodal points will be a complete waste of time if you fail to set foot on bare soil adjacent to any of them.

Arrive at a farm a few weeks too late and you'll probably miss the opportunity to search for an entire year. Arrive a couple of weeks too early and you'll find nothing but a field full of ripening produce.

I know two ways to acquire the essential knowledge. The first is to farm the land. Put yourself in a position where your economic survival depends on knowing when to plough, how deep to plough, how often to harrow, the time and temperature best for sowing, how to fight weeds,

Fig.7(b)

insects, the weather, when to harvest, and whether the yield is high enough to show a profit. The farmer already has all of that knowledge at his fingertips, but he is so busy struggling to earn a living he has no spare time for metal detecting as a leisure activity. You must pick his brains - and that's the second way to acquire the essential knowledge. However, you will have to approach him with care, and at a time when he can spare 10 minutes for a chat and he is amenable to silly questions from a townie.

Donkey's years ago, when I had a child at O-level education stage, she was "doing Geography". Her course included the study of regional agricultural differences. You know the sort of thing: Why is it better to grow wheat in

Fig.7(c)

Cambridgeshire than in Cornwall? Whenever I called at a farm on a quest for permissions I steered the conversation in the direction of children's homework and told the farmer my daughter was forever pestering me with questions about farm crops. Could I clarify a couple of points by pestering him about that field of peas growing down by the gate? It worked. Not only was the farmer profitably forthcoming about crops grown on his farm and in the neighbourhood, but my offspring passed in Geography.

Fig.7(d)

Here's another approach that has topicality and might appeal to those without young children. Take a trip - a bike ride is best - around the locality where you plan to concentrate your detecting activities. Visit each farm, telling farmers you hope to compile a list, perhaps for publication on the Internet, or in a parish magazine, naming all the vegetables, fruits, meats and other produce that people might buy from local growers. Ask about future cropping plans and when produce will be available for sale on each farm. You'll soon know everything you need to know about locally grown food, and I hope you will publish your findings. But what's more important is that you will have a

Fig.7(e)

very good idea where and when bare plough soil is likely to be available for detecting. Plan your permission seeking accordingly.

Don't forget that a change of crop involves a change in soil preparations. Some fields that fail to yield impressive numbers of finds in one season, may become highly productive once the farmer has ploughed before sowing deep-rooting vegetables.

(For quiz enthusiasts, Figs.7b., 7c., 7d., and 7e. show beans, cauliflowers, clover and maize. And just in case you failed to spot my trick question, Fig.7a. is wheat. If you thought it was barley you may find yourself arriving on a farm at the wrong time of year.)

8. Droving Routes

For nineteen hundred years of recorded history refrigeration did not exist. Every cow, pig, sheep, hen, goose - any creature raised for meat - went live to market, and it went most of the way on foot. When Queen Victoria began her reign in 1837, 5 million people lived and worked in London. They all wanted meat; the rich two or three times a day; the poor once or twice a week. The supply route began, as it had for centuries, in the west of Britain (Highland Scotland, Wales and the pasture counties of England) using a network of lanes, byways and bridleways that led not only to Smithfield in London's heart, but to numerous provincial markets en route.

At those nodal points some animals were slaughtered for local consumption; others joined larger herds, often hundreds, sometimes thousands strong....yes, just as you've seen them in Texan stampedes with John Wayne in the saddle....before setting off on the long drove to fattening pastures in Kent, Essex and Hertfordshire. It is recorded that in 1766 some 992,000 cattle (including 100,000 from Scotland) passed through Hertfordshire en route to Smithfield. They would have enjoyed recuperation for a couple of weeks on lush grass, putting back the weight lost during the long drove. A short journey to Smithfield followed, then slaughter and a butcher's block. In London their hides went by the million across the Thames to supply the leather industry in Bermondsey.

The movement of livestock on a similar, proportionate, scale occurred throughout Britain from Roman times until the Railway Age. Drove roads, also known as drifts, can often be recognized as wide verges alongside country lanes and footpaths. The verges provided some grazing for flocks and herds, as well as allowing wheeled traffic and horse-riders to use the road at the same time. The nodal points on droving routes will be where stopovers occurred, especially at crossroads and near old pubs. Nodal points that developed into markets witnessed the exchange of vast sums of money because all the dealing was done in cash.

My own experience of its richness comes from the Stagshaw Bank area in Northumberland where pastures north and south of Hadrian's Wall served as a huge open-air

Fig.7

market for cattle driven from all parts of Scotland. Millions of animals were sold around Stagshaw from Anglo-Saxon times down to the beginning of the 20th century. Thousands of hammered coins - both silver and gold - were recovered by the many detectorists who enjoyed the bonanza 20-odd years ago.

Those days may have gone, but many sites associated with droving remain undiscovered. Your county must have some associations with the trade: local markets; places where drove roads crossed; spots where herds forded rivers; encamping grounds; farms that were once in the livestock fattening business. Your first clue might be no more than a farm name as in Fig.7. A diligent search of the farm fields, especially former pastures now used as arable, might prove very rewarding. If you find plenty of iron horseshoes, or the smaller shoes with which cattle were shod prior to long droves, you have useful clues to where the animals rested overnight. Use those clues, plus what I've said elsewhere about water supplies, to locate spots where the drovers encamped overnight. That's where the money lies.

9. Stiles

A study of all the stiles in a single parish within your county would make a fascinating project guaranteed to produce some exciting finds, and one sure to interest the farmers whose permissions you requested. Your discoveries would span many centuries because stiles have been part of the British landscape since Anglo-Saxon times and some have probably existed on the same spots for just as long. Our word stile derives from the Old Saxon *stigen* meaning to climb, which is the very essence of the obstacle's usefulness: humans can climb stiles; cattle and sheep cannot.

As townsfolk making occasional forays into the countryside, we see stiles as oddly shaped stepladders that enable us to get over hedges and fences without hurting ourselves or damaging the barrier.

We forget that behind and ahead of every stile lies a right of way once so useful to the local economy or social life that somebody went to the trouble of ensuring that a hedge or fence to keep animals from straying did not impede a lawful right of way for local people. After detecting for several yards around any stile find out where the early footpath came from and led to. Your stile study will soon expand into a wider survey of local footpaths, especially converging routes.

A parish boasting lots of stiles will probably be engaged in mixed agriculture with different fields under plough at certain times of year, and with flocks and herds grazing and manuring different fields as part of the yearly rotations. Such farming methods offer detectorists the best chances of getting onto bare soil at various months of the year.

Here's another point worth considering if you've found a parish with lots of stiles. It must therefore have had lots of people clambering over stiles in the past. That will improve your chances of finds around all the stiles. It might also hint at some important local venue acting as a powerful nodal point that drew people over the stiles and along the footpaths.

Fig.9a. shows the basic step-over-the-fence design in use for centuries. The timber on much of this specimen is very modern. Nevertheless a Victorian OS map shows a stile at

Fig.9(a)

Fig.9(b)

this spot in 1875. If many people use it we can hardly be surprised that a stile requires frequent repair and renewal.

Fig.9b. shows a modern gate-type stile favoured by footpath preservation societies because mothers with go-anywhere pushchairs can enjoy rural rambles with only gate stiles to negotiate. Reference to early OS maps will tell you if the ground around a modern stile such as this holds Victorian or perhaps earlier losses.

10. Manor Farms

If I found myself in a totally unfamiliar English parish where a benevolent fairy godmother granted permission for one day's detecting on a single site of my choice, I'd go for a Manor Farm every time, provided, of course, that on the appointed day it offered bare soil on which to detect. The site would be easy enough to spot on an Explorer map where farm names are clearly printed. A sign similar to that shown in Fig.10. would confirm that I had arrived at the gate.

Fig.10

Follow my lead and you too will place yourself at the hub of your parish's prime agricultural land, surrounded by fields that offer the best local potential for detecting hammered coinage. Look closely at the earliest OS map of the area and plot the positions of all lanes, footpaths, springs, ponds and watercourses. If these features were shown on late 19th century maps it's highly likely that most of them go back to medieval times; even the Norman knight who came into possession of the manor after 1066 would probably have viewed a landscape not unlike that indicated on the Victorian map.

The following is a translation of a **Doomsday Book** entry typical of thousands more. (Note: the term "in demesne" means belonging to the manor.).

In demesne 5 ploughing oxen and 16 villeins with 10 ploughs; and a church there; and 7 serfs; and a mill worth 9 shillings profit; and ponds with 70 eels; and 20 acres of meadow; and woods for 30 swine.

No doubt the new lord and his successors extended boundaries, won more land from nearby virgin forests, improved drainage and built larger barns. But fundamentally, and for the next 900 years, the Manor Farm lay at the epicentre of all social and economic activity in the parish. Every gold and silver coin that came into the parish, apart from the silver halfpence and farthings minted in the early Middle Ages as a result of rising prosperity among farm labourers, probably passed through Manor Farm at least until the Church acquired a greater interest in money as a result of its own agricultural successes.

Even the local miller and blacksmith worked for the lord of the manor until the 14th century when many paid lump sums to gain independence. Similarly, the profits from all crops, fish, meat, wool, timber, indeed anything produced within the manor, went to the lord until toilers on the land began to pay cash rents to the landlord. Many of the rents were collected at meetings held on the Manor Farm, adding to opportunities for losing money on its footpaths and other nodal points.

In the reign of the Tudors open-field manors began to fade away. But by that time the soil on many of them was rich with hammered money and lost artefacts. One method of identifying places where coin losses must be greatest requires you to skim-read entries in the **Doomsday Book** for your part of England. (You can view this work in its entirety on the Internet). Look first for manors that had substantially more ox-teams and several more water mills than others around them. These will be the places where, in the 11th and 12th centuries, and probably later, the soil was highly fertile.

A test of continuing prosperity into the 14th, 15th, even 16th centuries will be passed if Victorian Ordnance Survey maps show evidence that moats were constructed around or close to what are now manor farms. The remnants of medieval moats will show on OS Explorer maps as three-sided, angular ponds, some still with a habitable dwelling within. Win search permission on the fields immediately flanking the moat site and your chances of medieval hammered will be very good indeed. Fields further out will probably still retain their fertility and be under plough. If you can spot telltale signs of medieval ridge-and-furrow, trace the lines of old footpaths between the blocks of ridge-and-furrow; that's where people and animals regularly walked and lost possessions. Any locations where old footpaths converge when crossing modern fields should also repay attention.

11. Ponds

The last time I drove on a motorway I made the stupid mistake of not checking my petrol gauge before joining a six-lane traffic jam called the M6. In consequence I had to suffer an embarrassing wait on the hard shoulder for an AA patrol carrying a spare gallon or two. It occurs to me now that petrol filling stations serve much the same purpose as pre-Victorian roadside ponds, even down to the similarity that at a motorway service station the humans refresh themselves in one place and the cars are tanked-up in another.

A couple of hundred years ago any drover or pack-mule driver would have made much better preparations for his journey than I did. Cattle and horses on a long trek needed to drink large amounts of water every fifteen miles or so.

Fig.11(a)

This requirement determined the chosen route because the track or pathway had to pass an easily accessible watering place roughly within that distance. Having no emergency services to fall back on, getting to the next drinking place became quite literally a matter of life or death for animals on the hoof.

One of the very few disadvantages of Explorer maps concerns their scale; it's too big to show features spaced at fifteen-mile intervals on a single sheet. Ponds (visible ones, that is) show up clearly enough in blue, but it requires two or more sheets aligned in order to spot that a cross-country route, perhaps now made up of minor roads, footpaths and bridleways, does indeed have strategically placed ponds along its length. The old one-inch-to-one-mile OS maps were small enough in scale, though too small to show anything but very large ponds. To investigate a string of ponds that once served a long drove route or packhorse trail you will have to spread several Explorer maps across your living-room floor. The effort might reward you with some very productive sites.

A drinking pond requires regular maintenance otherwise it will become weed choked. When hundreds of animals used it, stirring the muddy bottom around its perimeter, weeds could not get a foothold. Neglect results in what is shown in Fig.11a. In a few years time this pond may be nothing more than a ring of trees surrounding a grassy hollow.

Fig.11b. shows the effects of a lowering water table on what was a very large pond at the conjunction of three important medieval roads. Tree and shrub growth around its edges were avoided because the pond was used until very recently by travelling folk who held horse fairs on the adjoining common. Horses hooves rooted out any seedlings that sprang up. But as the photograph clearly shows, falling water levels are destroying this pond. On a Victorian 6in map it shows as a substantial stretch of water, but only in black and white because early OS maps were not coloured.

Time spent on transferring information about ponds shown on the old maps onto your Explorers should pay dividends. Use a blue marker pen. At the same time clearly

Fig.11(b)

record the Victorian positions of any small streams and springs in the immediate vicinity of the pond. Humans who drove or rode the animals would have used fresh water fairly close to the pond, probably encamping at the same place. A combination of a suspected drove road, a fairly large pond about 15 miles from another, a spring or brook nearby, and accessible ploughed land adjacent, indicates a nodal point with excellent finds potential.

Farm animals, especially plough oxen, also drank from ponds in ages past, when even the farmer and his family relied on springs and wells. Again we can count ourselves fortunate in having accurate 19th century maps that show the locations of even the smallest ponds. They are especially worth investigating when associated with ridge-and-furrow marks in the landscape, or with nearby deserted medieval villages, or with monastic remains. During ploughing of medieval fields the oxen had to be taken to water several times each day; their drinking ponds therefore became nodal points where coins and artefacts were often lost.

If monastic remains lie nearby any visible water might be the remains of fishponds. Sometimes a necklace of ponds fed by a single stream helped medieval monks to keep their fish stocks healthy.

12. Converging Routes

Take any Explorer map that covers an area of rural Britain. Spread it on your kitchen table, examine it carefully, circle every spot on the map where three or more public footpaths meet when crossing fields. Spend your next few detecting days winning permissions to search on those spots. If you don't make Victorian to modern finds on most of them, have your detector checked by your dealer. Something is wrong!

If all productive sites were marked as clearly as the bold green converging lines mentioned above life would be a lot simpler for the average detectorist, but I'm afraid you will have to carry out further research to identify converging route sites that hold ancient or medieval losses.

Here's the first complicating factor: many once-busy footpaths and bridleways have not come down to us as modern public footpaths. They may be marked on your Explorers as thin black dotted lines indicating either a private footpath or a parish boundary line that has fallen out of use as a footpath. So you can now go back to your Explorer and mark locations where one or two public footpaths meet one or more private paths or boundary lines. Win search permissions on those spots and your chances of earlier coins improve.

The second complicating factor: many ancient footpaths developed into modern roads. The busier a footpath was two or three centuries ago, the more likely its evolution to a bridleway, then to a highway. In many places 20th century tarmac now covers these old, even ancient routes. So another perusal of your Explorer is needed, this time seeking spots where modern roads are joined by several public footpaths, or private paths, or parish boundaries.

Fig.12a. shows the sort of complex convergence that can repay close attention. Several routes meet today at this spot. Three tarmac roads are visible; two more lie behind the central tree. On the right there's the beginning of a bridleway alongside an entrance to a private dwelling. This entrance path becomes a private track to fields at the back of the house. The island of grass and trees in the centre of the photograph has yielded Georgian and earlier finds in winter when the grass dies back, so it may have been used

Fig.12(a)

as a roadside fair site a couple of hundred years ago. The straight road has Roman foundations. In fact, it becomes an arrow-straight parish boundary just out of picture, and it is marked "Roman Road, Course Of" on the relevant Explorer map. Arable fields flanking this spot have produced Roman finds in the past; the private track alongside the dwelling leads directly to a place marked "Iron Age Pottery Found" on an 1870s 6in map. All of this evidence demonstrates the nodal importance of the area during 2,000 years of human comings and goings. Look for locations like it on your own maps.

Fig.12b. provides another perspective on converging routes. It's a snippet from a Victorian 6in OS map showing a village with seven or eight routes converging, and with a windmill and a blacksmith's shop nearby. Long, narrow gardens behind many of the dwellings point to a medieval settlement; such gardens were called tofts - patches of ground allotted to villagers for erecting small sheds and growing their own food. Win permission to search any such patches when the soil is bare and you could make interesting finds. Similarly, fields around the windmill will hold desirable coins and relics.

I would argue, however, that there's an even more interesting area in the lower left quarter of the map where

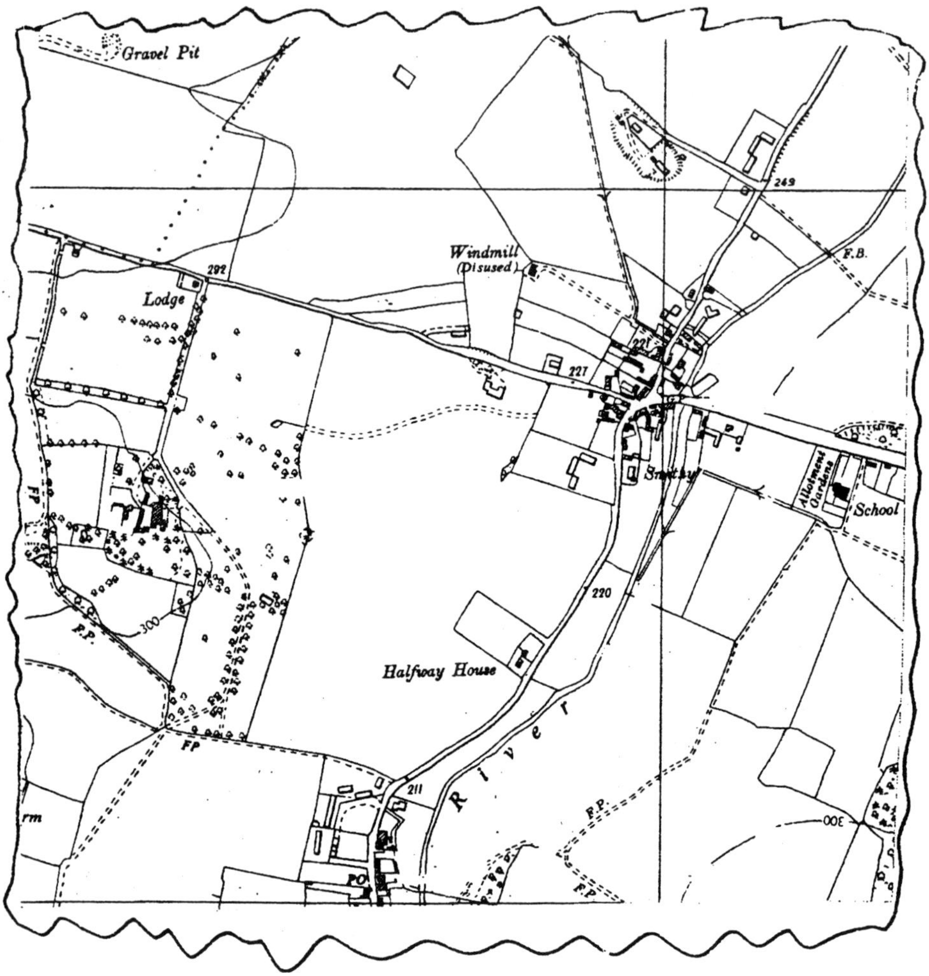

Fig.12(b)

we see the convergence of five footpaths, one of them leading down to the river and a probable fording place or the location of a vanished watermill. A large building, perhaps a former manor house, with its lodge clearly marking an important entrance gate, stands on rising ground that overlooks the village and the footpaths to the south. The lodge area and the possible ford or watermill both cry out for detector surveys, while the footpath nodal area could be the site of an early medieval market once held under the protection of a Norman motte and bailey on the hill. You could be at work on choice sites such as this if you put some time in at identifying converging route sites on your own local maps.

Fig.13(a)

13. Elevated Ground

The field on the skyline in Fig.13a. is always a bustle of activity in late July, weeks before crops are ready for harvesting in other fields on the farm. I asked the farmer why and he listed the following as important factors: The field gets the best of the light, thanks to its open aspect. The soil is freer draining than soil in lower fields, so it never becomes too wet even when rainfall is high in spring and early summer, as in recent years. The field also benefits from plenty of sunshine because the plateau has a slight slope towards the south and the sun is in the southern sky during late summer months when crops are ripening.

It occurred to me that a Victorian farmer would probably have given the same reply to my question because the light, the natural drainage and the late summer sunshine were almost identical in the 19th century. I could say similar things about medieval, Norman, Anglo-Saxon, Roman, and even Celtic farmers. They all preferred fields that offered those benefits and gave the resulting higher yields.

Does that mean that every farm where you encounter elevated arable fields with good drainage and a southern

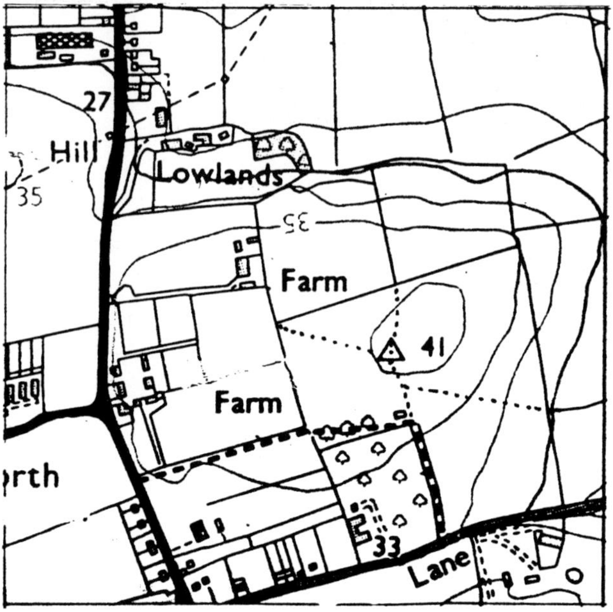

Fig.13(b)

aspect might have pre-Victorian coins and artefacts in its soil? No. Two crucial questions demand answers before you can consider that possibility. The first: When was the land cleared from the primeval forest and put to agricultural use? Archaeological records might answer that question. If they confirm ancient structures or report chance finds nearby, it's a fair bet that the higher ground was farmed in ancient times, though large-scale agricultural expansion did not get underway until the early Middle Ages. The second question: When did crop yields from these fields, and the neighbouring land, exceed the needs of local people? Lots of old footpaths and tracks crossing the land hint at workers, livestock and produce on their way to markets. If some of the footpaths were incorporated into ancient local boundaries, those elevated fields might hold more than high yields of crops. You can identify fields with all of those attributes in your own county. Look at illustration Fig.13b. It shows the contour pattern and converging footpaths layout you should look for on Explorer maps. Ploughed land around spot height 41m in my example should hold interesting finds, perhaps pre-Victorian.

There is much low-lying arable land in Britain, a lot of it converted from a waterlogged state into rich crop-growing hectares thanks to Victorian and modern drainage. But in medieval times fen and marsh persisted around most river estuaries, and in wide swathes of East Anglia, Somerset, South Lancashire, the Upper Forth, Stirlingshire and Fife.

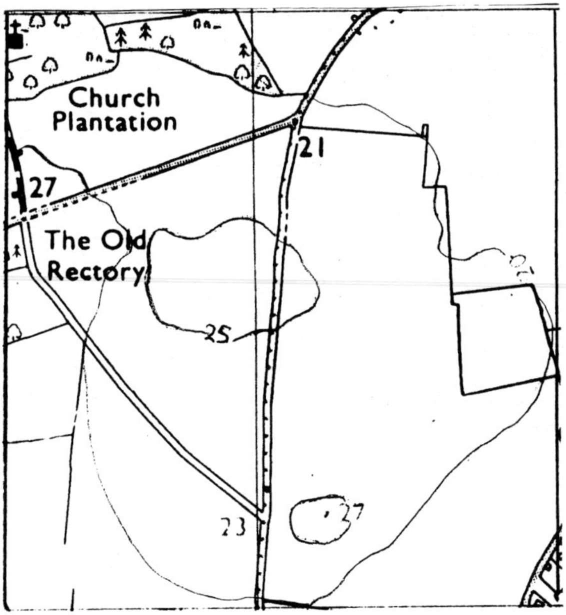

Fig.13(c)

Nevertheless, these areas enjoyed prosperity with a way of life based on wildfowling, fishing and the exploitation of rich summer pastures. If you search arable land in those regions today you should first examine old 6in OS maps, looking for slightly elevated ground. Even if you can see no old footpaths hinting at nodal points, these relatively high spots probably attracted people in the past. Some may have been island habitation sites in medieval times, reached only by small boats that left no footpath evidence even though the islands were clearly nodal. Fig.13c, showing part of a 1920s 6in map, offers an example. Note the wide spacings between contours, indicating fairly level ground. Most of the land has a height of 20-23ft, but there are two areas where the contours indicate 25 and 27ft. If they were islands in the distant past, a detector search, with the aid of a GPS positioner, along the line of the 25 then the 27ft contour, might prove rewarding. You could be searching on former island foreshores that drew many people to their banks.

Let me add here that whenever you work in or walk on a harrowed field that seems at first glance to be almost bowling-green flat, kneel down and look across the soil from near ground level and in all directions. Patches of slightly raised ground (no more than a few inches higher) may reveal themselves. Always detect around their peripheries; they might have been habitation sites in the distant past, their levels raised over centuries by accumulated debris and by recent draining of adjacent swamps.

Fig.14(a)

14. Churchyard Walls

I explained in the Introduction why churchyard walls qualify as nodal points of communication: grass roots or soil along the outside of the wall can hold Anglo-Saxon to Georgian coins and artefacts lost by people who used the wall as a back-rest in ages past. A vicar's permission to search the sort of wall shown in Fig.14a. should prove profitable. The brickwork is Victorian or earlier and a notice on the church door states that there was a church on the site in the 14th century. The wall must have been rebuilt, but its foundations indicated to the bricklayers precisely where to build the new one, so earlier losses must still await discovery.

Fig.14(b)

Fig.14b. shows what has happened to many churchyard walls over recent decades - they have fallen due to neglect and lack of funds to rebuild. So here we have a church dating from Anglo-Saxon times surrounded by a sparse wooden fence. The ancient yews alongside it look quite incongruous growing next to modern timber, but they do confirm the line of the original wall. A detector search might locate quite a few nails dropped by the fencing contractors, but other signals could come from Anglo-Saxon and medieval losses.

Fig.15(a)

15. Hollow-Ways & Hedgebanks

As a cyclist I'm highly sensitive to road gradients. The slightest up-hill slope transmits its resistance to my tyres and leg muscles within a few turns of the wheels. Perhaps that's why I'm so aware of stretches of road which look like cuttings designed to reduce gradients on hills, but which don't make my leg muscles ache because the slope remains fairly level even when the earth on both sides of the road rises to shoulder height as in Fig.15a.

At such locations I usually jump off my bike, scramble up the sloping verge and hope that I'll find arable fields flanking the road. If I'm close to an old village, or if there's a medieval church nearby, I've probably parked my bike in a hollow-way, which may be the modern vestiges of an ancient track that once served an early settlement, allowing livestock to be driven along it with no danger of them eating crops in adjacent fields. Centuries of use carved it deeply into the landscape; later road improvements made it even deeper when engineers laid kerbstones to support the steep sides. Permission to search the flanking fields often produces early finds.

If you discover such a hollow-way follow its line until

Fig.15(b)

you reach a spot where an apparent gap occurs. That might be the site of a lost gateway, or the location of a busy local crossroads centuries ago.

In wooded countryside similar banks are often encountered - linear mounds of soil with shrubs and small trees growing out of them. Some are sited quite close to roads, but others occur within woods and a long way from modern roads. Fig.15b. shows one within a wood, the shot taken looking out of the wood towards arable fields that can be glimpsed through the foliage. From this position the purpose of this hedgebank seems clear: it was made to deter deer from leaping out of the wood into growing crops. Again, you should seek gaps where humans controlled deer movements and perhaps lost possessions. Other hedgebanks once served as property boundaries. Ploughing obliterates them on open fields, but remnants linger within woods. Try extending the line of a woodland hedgebank on your map. Does it point towards a deserted village or a vanished dwelling? Does it align with local footpaths and converge on a nodal communication point?

Fig.16(a)

16. Nettles & Elders

The common stinging nettle (*Urtica Dioica* is its Latin name) shown in Fig.16a., "thrives on disturbed ground" according to botanical textbooks. The same is often said of the common elder shrub (*Sambucus Nigra*) shown in Fig.16b., because both of these plants love soil that has very high levels of nitrogen. None of the experts' textbooks I have thumbed through go on to say that the cause of high nitrogen levels is almost always human activity, especially depositing waste materials in the ground. As the waste breaks down the soil becomes highly enriched with all the nutrients plants love best, nitrogen being one of them.

Look closely at those weeds growing as crop marks and indicating the positions of former hedgerows. Most of the weeds will be nettles. The farmer will uproot them before they can grow to any great height, except on set-aside fields where nettles along old hedgerow lines can reach three feet before the fields get ploughed in autumn. Elder seedlings don't have much chance in that situation because they don't germinate until the spring of the following year.

A tall and vigorous patch of nettles covering several square yards and growing on a grassed field in June or July almost always points to buried refuse. If you also spot flat-topped sprays of white flowers growing on a gnarled shrub

Fig.16(b)

nearby (in September the flowers become black berries) you've found an elder and you have almost certainly discovered a buried refuse pit. It may be a recent deposit. Ask the farmer. If he says nettles and elders have grown on that spot for many years ask for permission to carry out a trial dig. You might discover delightful glass and ceramic bottles and containers. They were discarded as unwanted refuse about a hundred years ago. Today they are highly prized by bottle collectors. Tens, perhaps hundreds of thousands, have been recovered from the ground during the past 30 years. The sites they were found on - old chalk pits, former quarries and especially the sites of Victorian brickworks where household refuse was once sorted to extract coal ashes to mix with clay - had one thing in common: nettles and elders grew profusely on them, alerting bottle collectors to the possibility of exciting finds. Although many of the spectacular sites were excavated decades ago, there must be thousands of small dumps still undiscovered. Concentrate on finding nettles and elders and you might soon turn to bottle collecting as an additional hobby. Victorian 6in maps show the locations of large numbers of pits, quarries and other interesting holes in the ground. They also mark the sites of brickworks. A search of such places in your parish is the way to start.

17. Trees & Hedgerow Shrubs

It is hardly surprising that the towering oak shown as Fig.17a., standing in majestic isolation (the background trees line a driveway to the farmhouse) in the middle of a field of wheat, caught my eye as I cycled past. It took two or three hundred years to reach a height I estimate at around 100ft and it must have attracted many people in the past. I suspect that it's a remnant from a long-vanished hedgerow, though no hedge is shown passing that spot on the Victorian OS map I consulted.

However, it was common practice to allow a few saplings to grow to maturity along boundary hedgerows in the past. In leaf they provided shelter for animals, as well as something to nibble at on their lower branches. Decades later they gave a valuable supply of timber for building purposes and for furniture making. Oak was always the most widely grown, but elm, ash, willow and poplar were all useful in hedgerows.

In centuries past everyone knew the precise locations of the largest trees in the parish. They were popular meeting places for lovers, children and herdsmen. Together with shrubs growing along the same lane or footpath, a mixed hedge provided a sustainable supply of nuts, herbs, fruits, animal feed and many other needs of the community. Although hedgerows cannot be accurately dated by counting specimens, it's worth learning how to recognise certain trees and shrubs, if only to appreciate why the verges alongside species-rich hedgerows must have witnessed plenty of human activity in the past. Here's a list of useful hedgerow trees and shrubs known to have grown in medieval hedgerows:-

Ash - hardwood used in curved furniture making because it bends readily after steaming; poles used for farm tool handles.
Beech - nuts widely used as fodder for pigs.
Blackthorn - its bark was used in tanning, and in making black ink; its fruits (sloes) were made into preserves, and to flavour gin; its straight stems were ideal for walking sticks.
Bramble - fruits for pies and jams; stems used as ties for thatch and other bundles.

Fig.17(a)

Buckthorn - berries used as a purgative.
Crab apple - fruits used in jelly making.
Dogwood - berries yield an oil formerly used in lamps and in soap making.
Elder - blossoms used to make wine and cordials; berries used to make wine and in cookery; branches in bloom tied to harnesses as insect repellent; immature wood used to make children's toy whistles; mature wood ideal for small tool handles; bark and leaves used in dyeing.
Elm - once the commonest tree in the English landscape, now only dead stumps remain, due to Dutch Elm disease; used for longbows, coffins, water pipes and cartwheels; foliage makes good cattle fodder.
Hawthorn - probably one of the first shrubs used by early farmers to construct hedges; its name, in Old English (*haga*), means hedge; it can live for 500 years or more; in Celtic mythology a magical plant; in medieval times blossoms used to decorate maypoles and May Queens.
Hazel - nuts as winter food; thin stems used in wattle-and-daub building and in hurdle making; thicker stems used to hold down thatch; thickest stems used as walking sticks.
Holly - important in both pagan and Christian rituals; leaves a useful winter fodder for cattle; when boiled twigs yield bird lime, a sticky substance formerly used to catch small birds such as finches; timber used in furniture making.
Hornbeam - extremely hard wood ideal for chopping blocks, cogs in water/windmills, ox yokes; coppiced as a source of fuel.
Oak - extensively used to build houses, barns, churches and ships; pollards yielded vast quantities of wood for charcoal making; acorns fed to pigs.
Rose - petals were eaten or used as a perfume; rose hips, which have 20 times as much vitamin C as oranges, were made into medicinal cordials.
Rowan - important in pagan ritual; berries were collected for use as bait in bird traps; berries also used to make a vitamin C-rich jelly, eaten as a scurvy preventative.
Spindle - berries were dried, crushed and used as insecticide; boiled berries used as hair wash; other parts

Fig.17(b)

used as dye; fine-grained wood ideal for making spindle-whorls; skewers and knitting needles.
Sycamore - introduced to Britain in Tudor times and prized in furniture making.
Wayfaring tree - said to grow in hedgerows flanking old droving routes; stems used to make rope.
Whitebeam - widely used in Anglo-Saxon times as a marker tree on boundaries, thanks to its distinctive white foliage.
Wild privet - leaves used as poultice for abscesses and ulcers.

Fig.17(c)

Even in our electronic age gathering hedgerow fruits and nuts has not totally died out as a summer and early autumn pastime. The ancient beech tree shown in Fig.17b. attracts scores of local children towards the end of the summer holidays when its branches are festooned with nuts. A local organic pig farmer buys the youngsters' nut harvests. There is stiff competition between the kids and neighbourhood grey squirrels. However, it's the children's lost coppers and trinkets that litter the ground beneath the old tree. Similar losses must have occurred in times past along other hedgerows.

The upper classes were never content with English nuts. From Tudor times country landowners often planted exotic

Fig.17(d)

trees in their private parks and gardens. If the owners or heirs fell on hard times and the parks and gardens reverted to agricultural use some of the ornamental trees survived in the landscape. Fig.17c. shows a sweet chestnut that probably started life as a sapling on a rich estate. My folding bike alongside gives an indication of the girth of this specimen. Who knows what choice trinkets lie beneath the surrounding grass?

Trees were also grown to maturity in local woods. If you encounter an aged giant like that in Fig.17d. you will have bare, detectable ground beneath your feet even at the height of summer; the giant's leafy canopy blocks light and prevents other plants and shrubs thriving on the ground beneath its branches. Copper coins in such ground conditions will not survive, but if gold or silver coins and possessions were lost they will still be worth finding.

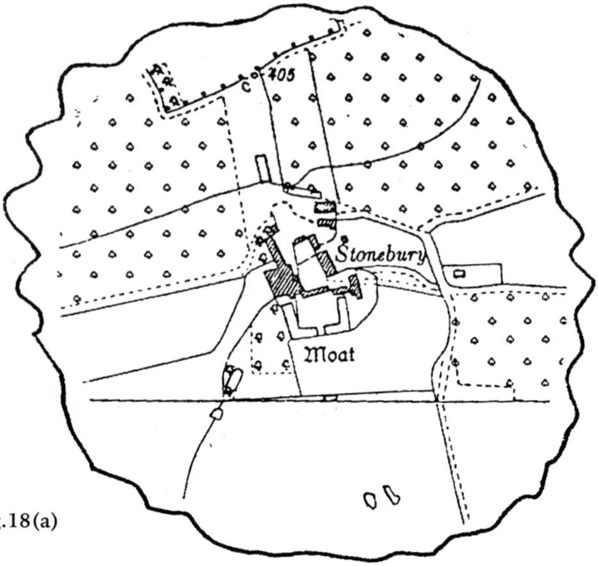

Fig.18(a)

18. Moated Sites

Explorer maps of southern and eastern England have a generous peppering of moats, clearly marked with that very word. It also appears on maps covering other parts of Britain, although less frequently. Most of the southern and eastern moats were created during the Tudor Age. You might wonder, as I once did, why moats achieved such widespread use at a time of relative peace and quiet in the English countryside. We all have a mental image of a moat and a spiked portcullis as a means of keeping enemies at bay and possessions safe behind a raised drawbridge. In fact, lowland moats were the Tudor equivalents of modern double garages, barbecue patios and decking: status symbols proudly displayed by yeoman farmers grown prosperous thanks to their skills at crop raising, livestock breeding and getting good prices on market days.

Farm incomes doubled between 1500 and 1600, thanks in part to Henry VIII's take-over and subsequent sale of Church land, most of it bought by young men eager to make their fortunes in agriculture. As the profits rolled in the nouveau riche began to emulate the landed aristocrats by modernising their farmhouses, then surrounding the new buildings with a neat moat. Today the farmhouses are often the homes of successful City commuters, while the moats have largely dried up to leave one or more oddly-shaped ponds close to the house. Modern owners will know something of the history of their property, but they may have few relics of the farmhouse's early years. Win

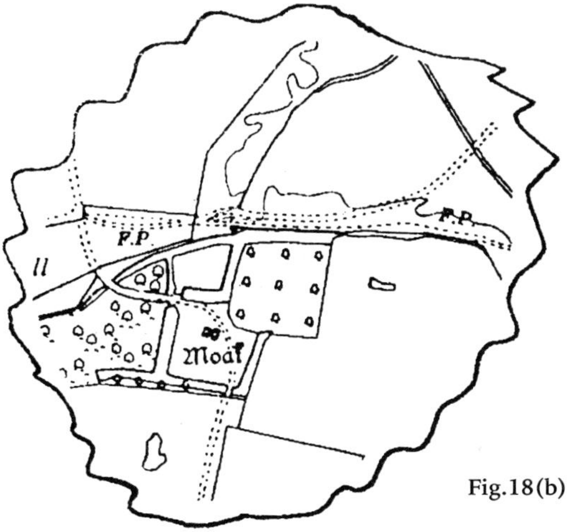

Fig.18(b)

permission to search gardens, even to clean out the bottom of the moat, on a share of finds basis and your efforts should be handsomely rewarded.

Figs.18a. and 18b. show what you will typically see on your map: narrow pond-like stretches of water, usually exhibiting at least one right-angled bend. Traces of the complete moat may be visible as depressions in gardens and vegetable plots adjacent to the house.

The moat was but one manifestation of Tudor wealth, an indication that an upwardly mobile 16th century family was spending lavishly on personal possessions such as jewellery, silver shoe buckles, and ornate buttons for best clothing. Household servants, emulating their employer, would also have had their cheaper baubles and beads. And the farm labourers, perhaps marginally better paid than those who worked on neighbouring farms, would have had a few extra silver coins to rattle in purses and pockets. The fertile fields around the moat would inevitably have become further enriched as coins and artefacts found their way into the soil, the losses occurring at footpath junctions, the bases of large trees, stiles, and at other nodal points.

In borderland regions of Scotland and Wales moats were dug for genuinely defensive reasons. Their modern sites are probably just as rich in coins and relics, but dairy or sheep farming may still be the local way of life, offering detectorists and field walkers poor prospects for getting onto ploughed fields, although you should still try to gain permission for searches of vegetable gardens at the rear of the house. Ask the present-day farmer or occupant of the dwelling if there are plans to abandon sheep or dairy farming and to grow arable crops in the adjacent fields. When that happens expect large numbers of fascinating finds.

19. Eyes-Only Techniques

Recognising the textures, colours and designs on glass and ceramic objects from all ages, Celtic to Victorian, presents few difficulties now that digital cameras, video recorders, and even old-fashioned colour films, offer such high standards of performance. You can record images or take non-flash photos for personal use almost anywhere in the British Museum, where you will find amazing displays of ancient and old pottery, glass and building materials in the cabinets. If you can't travel to the British Museum, try searching the World Wide Web. The keywords pottery, glass and archaeology will soon have your screen full of close-up shots depicting everyday wares and building materials discovered during excavations all over Britain. Print out what you need as a reference collection, but please remember that there is no substitute for holding the real thing - which you will not be allowed to do at the British Museum.

It's important that you develop an eye for non-metallic surface finds because they are indicators of nearby nodal points, usually some form of habitation site, or places where large numbers of people came together and dropped lots of litter.

On the next rainy day on which you can be sure of encountering newly ploughed or harrowed soil in your general search areas, get out and about with a small collecting bag and your binoculars. Try to position yourself in the fields so that the best light comes from over your shoulder. Walk the soil, eyes focused a few yards ahead of your feet, and pick up any fragments of pottery, glass, stone, brick and tile that catch your eye. On some fields you'll fill your bag in no time, so discard duplicates on the field, or in a corner of the field. Don't make the mistake of putting unwanted items from your finds bag onto other patches of ground; you may be rediscovering them in a few years!

Standing on stiles, or even shinning up trees, might give you a broader view of the field when you scan with a pair of binoculars, but as I've had several tumbles from such wet perches let me warn you to take care. Look especially for patches of ground where surface clutter seems concentrated. Search those spots before you begin general field walking operations. If the clutter contains plenty of old fragments an immediate detector search in the vicinity might bring early successes.

You will, by now, appreciate that what you find is largely determined by the layout of nearby footpaths and the locations of nodal points. Take those into consideration when deciding which fields you will subject to intensive eyes-only searches. Also bear in mind that a single deep ploughing of any field will bring up thousands of fragments (even coins and artefacts) that have not seen the light of day since the last time a deep-rooted crop was grown. Your own notes on local history will have identified possible nodal points, some of which may not have yielded good finds so far. Deep ploughing will surely alter that, so eyes-only searches on all potentially productive fields are worth repeating between each new crop sowing.

I recently re-read a fascinating archaeological report written by Ivor Noel Hume, the author who gave us that wonderful book **Treasure in the Thames** way back in the 1950s. He went to live in the USA in later life and took to American archaeology like a Thames swan to water. The report dealt with Hume's excavations on sites in Virginia dating from the time when British colonists farmed the land. He explained that tobacco growing soon depleted soil of its nutrients, and that every three or four years each plantation owner moved his operations lock, stock and barrel to virgin land half a mile deeper inland where a new plantation was carved out of the primeval wilderness. Among the planter's equipment was a meagre supply of large iron nails used to join the rough-sawn planks that made the farmhouse, the tobacco drying sheds and the slaves' huts. They were so precious that every farmer recovered those he used on the old plantation by burning down the abandoned buildings. His slaves sieved the ashes to recover the nails. Hume found many slightly elevated patches of ash-rich soil; but he never found iron nails. To date the sites he relied almost entirely on pottery and glass fragments that came to light on the edges of the ash patches. The finds dated the buildings to 1590-1620.

Celtic and early Anglo-Saxon farmers worked in a similar

way; they too abandoned a farmstead when its soil's fertility diminished and crop yields began to fall. Perhaps they didn't burn their timber buildings, although all traces of many Celtic and Anglo-Saxon farmsteads have vanished. Luckily the families left plenty of broken ceramics in the adjacent soil. If you want to find their non-ferrous losses, improve your Celtic and Anglo-Saxon pottery recognition skills. Evidence of larger Anglo-Saxon settlements now vanished from the landscape can also come to light as a result of pottery finds. It was common practice in those days, as now, to bury the dead in particular locations. Often this involved burying a large pot as part of each inhumation. Centuries later ploughing may have disturbed and broken it and left fragments in the soil. Ploughed-out burial mounds also left broken pottery clues to their locations.

Prosperous Romano-Britons, and richer medieval farmers about 800 years later, built with brick and tile. Both had their own distinctive pottery and glass, but the most important difference between the two societies is that by late-medieval times farmers had learned much about how to maintain soil fertility by crop rotation, by the use of animal manure, and by allowing fields to lie fallow at certain times. Good farmland was no longer abandoned; it was acquired and improved, its boundaries clearly defined, its ownership handed down from father to son. What binds the late medieval agriculturalists to their Celtic and Anglo-Saxon predecessors is the very large numbers of ancient routes, boundary markers, watercourses and natural landscape features such as hilltops that were known and used by all of them during 1500 years.

Victorian farming didn't change beyond all recognition, despite the gradual introduction of machinery. However, from a fieldwalker's point of view one major change occurred: farmers began to spread on their ploughed fields large quantities of fertiliser-rich night soil (a blend of fire ashes, chamber pot contents and unwanted or accidentally broken pottery and glass fragments thrown out with other household refuse). Only the pottery and glass fragments survive to this day. You'll spot them during your fieldwalking activities. Large quantities indicate fields that were intensively manured. They must have been ploughed once a year by teams of workers and animals, so footpaths leading to them merit careful searches.

Let me now summarise what you can expect to find during eyes-only fieldwalking on freshly ploughed and/or harrowed fields where other evidence suggests human activity during past eras:-

Celtic. No bricks or tiles. Coarse black, brown, grey pottery showing no sign of turning on a wheel; sometimes with incised geometric shapes pressed into the clay.

Romano-British. Pottery scatters very common near villas and native farmsteads. (For example, in 2nd and 3rd century Northants there were two farmsteads in every square kilometre). Roofing and hypocaust tile fragments, reddish in colour, with no wavy edges, some pieces with square section, often found near villa sites. Well-made, sometimes finely decorated pottery fragments likely. Thin-walled, wheel-turned, colour-coated pottery common. Samian ware, which is uniformly red or black, polished, with no indentations on its inner surfaces because it was press-moulded rather than wheel-turned, may turn up on richer villa sites. Cubes of soft, reddish stone about the size of two Oxo cubes likely to have come from a domestic mosaic pavement. Smaller cubes, brighter colours, harder stone, suggest a mosaic from a villa. Deep blue, chunky glass may be Roman; fragments from long-necked, thin, green bottles may be Roman.

Anglo-Saxon. Stamped decoration on thick, whitish and grey pottery.

Medieval. More than 3,000 deserted medieval villages have already been recorded in England alone. Others, especially hamlets and individual farmsteads, await discovery. Timber building was still the norm (except for the church) but bricks commonly used for baking ovens. Glazed floor tile fragments, possibly geometrically decorated in yellow, brown, green, and about three times the thickness of modern floor tiles, would indicate a substantial building. Thinner tiles might be medieval roof tiles. Pottery will be

Fig.19(a)

coarse, gritty, thick-walled, mottled green, yellow or brown glazed. On monastic sites expect more worked stone showing chisel marks; more glazed tile fragments; more lead in the ground.

Tudor. Brick fairly common. Decorative detail apparent on some stonework. Lead fragments common near larger dwellings. Salt-glazed pottery fragments with speckled surfaces often found.

Stuart/Jacobean. Smudgy blue designs on thickish white pot might be Dutch or English delftware.

Georgian. Window glass fragments likely. Insignia of local landowners often embossed on bricks. Ornate cast-iron fragments from fire grates, foot scrapers, etc likely finds. In northern regions brick used for more farm buildings such as dovecots and corn-driers. Pottery bottles fairly common. Thick, black glass from broken porter bottles. In southern England knapped (= flat-faced) flint might indicate demolished walls.

Victorian. Widespread use of glass bottles and jars, usually embossed with trademarks, etc. Tobacco pipe stems ubiquitous on fields spread with night soil. Red and yellow clay drainage pipes now used in field drains. Wavy-edged red roof tiles common. Welsh slates also in use.

Fig.19(b)

Figs.19a. and 19b. show collections of stone, brick, glass and ceramic fragments picked up on two fields where arable farming has continued for many centuries. The first field has documentary evidence of possible Roman habitation, and of a medieval hamlet for which there was documentary evidence circa 1400, but the hamlet was not recorded on the earliest OS maps. Finds from that field (Fig.19a.) included a piece of what might be Roman hypocaust tile; a fairly thick pottery fragment which could certainly be medieval; and oyster shells, suggesting former human habitation nearby. Detecting on this field should produce ancient up to late-medieval coins and artefacts.

Fig.19b. shows surface finds indicative of night soil dumping in the 19th century. They include Victorian slate fragments; thin crockery shards with colour-printed decoration; moulded glass; machine-made bottle-tops; a machine-made wine bottle base; pottery jars with very accurately turned shoulders and bases; and many clay tobacco pipe stem fragments. Detecting on this field will probably yield large numbers of Victorian brass buttons, Victorian copper coins, and perhaps 19th century army badges.

20. Reading Maps

I can think of nothing more pleasurable than relaxing in an armchair on a wet afternoon in June with a spellbinding OS Explorer map in one hand and a fine-line pen or a highlighter marker in the other. I feel no shame in calling myself an armchair treasure hunter at that time of year when there's no bare soil to swing a detector, especially if the tides are out of sync for beach or river work.

My precious leisure hours could not be more productively spent than by perusing a map to find new sites for the coming season. And at times it can be as exciting as reading detective novels. There are clues to look for and decipher; lines of enquiries to follow, and if I solve the mystery it eventually leads to the recovery of substantial amounts of lost property. I may not be able to return it to its rightful owner, but I can share it (or any rewards) with grateful landowners. Let me tell you a little more about how I approach the task.

If the map depicts agricultural land, the first things I look at are the contour lines. They've been marked on all large-scale OS maps since Victorian times and to a trained eye they offer the equivalent of a three-dimensional view of the countryside. (If contour lines are entirely new to you, they are those very fine lines in the background of the map. Each line joins up spots on the ground that are the same height above sea level. Lots of contour lines very close together indicates that you are looking at a very steep slope; when the contours are far apart they tell you that the land is relatively flat.) The top of each map indicates north (approximately) so it's easy to pick out south-facing slopes because the contour heights will be increasing in a northerly direction. Those slopes were eagerly sought by ancient (and modern) farmers, especially in areas where the soil is fertile and easy to plough. If the map is showing arable land I usually shade-in all south-facing slopes the first time I browse a new purchase.

Next I look at the drainage pattern. Tiny threads of blue on the map indicate small streams. Each flows into a broader stream, which in turn flows into a river on its way to the sea. It takes quite a feat of imagination to appreciate that in centuries past each of those tiny threads would have been a substantial stream or rivulet, perhaps used by two- or three-man cargo carrying vessels, or used to power a vanished watermill, or used as an important source of drinking water for animals, or even humans. If other factors (nearby ancient habitation sites, for example) warrant it, I sometimes use a blue highlighter pen to emphasize the location and flow of local watercourses.

My third task involves marking all ancient and historical routes, especially Roman roads, pilgrims' paths, medieval highways and any additional information about local communications I might have come across during book research. Again, a highlighter pen contrasting with the Explorer's colours is used to remind me of these important clues. With that done I can now transfer to my Explorer everything I've learned about Celtic, Roman, Anglo-Saxon, Norman, Tudor, Civil War, Georgian and Victorian use of the land and its resources.

My maps used to have a jumble of notes scrawled across them until I devised a simple coding system that combines the first letter of the historical period with a number to indicate different types of sites, finds, etc. Thus the code AS3 neatly written in tiny letters using a fine-line pen tells me that an Anglo-Saxon coin was found close to that spot. I'm in the process of transferring reams of handwritten notes to my computer and I've used parish names to provide an alphabetical gazetteer. Here's a fictitious entry to show how it works: LAXTON, Northumb. AS3. 1955. Summersuckle Farm. NZ1234 5678. A quick cross-reference between map and computerised notes tells me the full story of what happened on Summersuckle Farm way back in 1955, and more importantly the exact spot where the coin find was made. A GPS positioner would get me to within a few metres of that precise location should I need to show the present farmer where the discovery occurred and why he should let me and my detector sweep the ground in hope of more to come.

Of course, an Explorer is never fully marked up. Even on a map peppered with codes telling me the locations of windmills, watermills, ancient fords, bridges, Anglo-Saxon churches and cemeteries, Tudor moats, Norman castle mounds, Roman villas, roads and military sites, Celtic farmsteads, ploughman's hoard spots, crop marks, and a hundred and one fieldwalker's eyes-only finds, there are always more to come. The marked-up map is a *vade mecum* that I can carry on every trip to a local reference library or county archive, ready to receive additional information gleaned from old maps and records available for inspection in those establishments. In this respect a marked-up Explorer differs from a detective novel: it's a better read every time I open it.

If I could afford the prohibitive costs involved, I'd always buy a large-scale aerial photograph of any farm where I hoped to persuade the farmer to grant fieldwalking, then detecting permissions. I've yet to meet a farmer who wasn't captivated by the view of his farm from about 5,000 feet. If the photograph indicates crop marks in arable fields the farmer will often ask you, before you can make the request, if you'll come back and detect the fields immediately after harvest. And if you could afford the cost of it, a present of the aerial photo would virtually guarantee search permissions granted for several years ahead. Well, you may not be able to afford aerial photos, but an Explorer makes a fine substitute. It would cost you little more than a fiver to present the farmer with a fully marked-up version, including all your notes relevant to his farm transferred onto it. I'm sure such an offering would instantly win you a new farm to work on.

21. Boundary Lines

Let's take a boundary line we all still recognise, even though it lost any importance as a demarcation between properties more than a thousand years ago: Hadrian's Wall. It told the known world where the Roman Empire ended and the barbarians began. At the same time it proved a magnet for the very people it attempted to keep out. Tribes from what we now call Scotland brought trade goods (gold, pearls, horses, slaves, semi-precious gemstones and more) to various gates and customs control points along the Wall where markets sprang up.

The tracks that led to those trading spots became the highways of later centuries. At least two of the markets survived down to the end of the 19th century. Small settlements developed at other trading points, but they withered when the border moved, first south, then north, in Anglo-Saxon and medieval times.

Let's look at another boundary line. Use a powerful magnifying glass to inspect one of your Explorer maps and make quite sure you can distinguish between the marks used to indicate footpaths (public and private) and the very different marks used to indicate parish boundaries. It may surprise you to learn that most parish boundaries have not altered since Anglo-Saxon times, just a few hundred years after the Romans left and Hadrian's Wall was overrun by barbarians.

During more than a thousand years people living on either side of a parish boundary have had good reasons to keep the boundary fixed. The first was, of course, to protect ownership of the land which lords of manors always hoped to hand-on intact to their heirs. Secondly, two parish priests claimed 10% of all crops grown on each side of the boundary; they resented giving up so much as the width of a footpath in an arable field. And each thought any footpath should run on the other parish's side of the boundary.

Disputes frequently flared-up and became noisy meetings along the boundary to sort out conflicting claims. When people meet regularly at a particular place somebody will soon decide to sell cakes and ale. Thus are local markets born. Thirdly, upholders of law and order (beadles, watchmen, constables and their ilk) were always keen to

Fig.21

make sure troublemakers and anyone who might prove a burden on local taxpayers stayed on the other side of the parish boundary. Their concerns and actions made all local people fully aware of the parish boundary. That's why they frequently met along its length. Occasionally they met at such a spot to avoid the clutches of the law in one parish by stepping over the line into the next. Thus were the locations for duels, cockfights, bare-knuckle boxing matches, unlicensed markets, non-orthodox religious meetings, and outlawed political rallies decided upon. Find two or more footpaths converging on a parish boundary and you have a spot where people congregated in the past. Find a location where three parish boundaries meet and you have another place that must have drawn people like a magnet in the past. Arable land nearby will be rich with losses.

The places where parish boundaries make sudden and sharp turns should always be visited and examined closely. When a boundary was first set, probably in late Anglo-Saxon times, although possibly even earlier, there would have been an obvious reason for making a sharp turn: an ancient burial mound, a very large oak tree, the summit of a hill, a ruined building, a spring, a place where rivers or streams met....something that made the precise location well-known. You may see nothing obvious when you view the place, but evidence of its importance as a nodal point will lie in adjacent soil.

Fig.21. shows a parish boundary. It is also the only surviving footpath now dividing two huge arable fields. Using careful measurements and comparisons between old and modern OS maps we can work out the positions of all vanished footpaths nearby because the ancient parish boundary provides an unchanging reference point. Note that even today one of the landowners has felt it necessary to cut the remaining hedgerow bushes in a distinctive manner so that they emphasise the boundary line.

22. Isolated Churches

Christianity's grip on the British psyche has weakened to the extent that in most cities and towns church services are poorly attended. What a contrast with our behaviour a century ago when congregations filled every church throughout the land to overflowing three times on Sundays, with most folk putting in an appearance once on weekdays too! Fear of retribution from the parish priest as much as from an omniscient god often drove less pious individuals to the church door. Even in the early 20th century there were parts of Scotland, Wales and most of Ireland where anyone who failed to turn up on Sunday was likely to find himself and his family barred from local shops and from employment for the rest of the week.

So, when you look at the church in Fig.22., which stands today in the middle of a one-acre orchard surrounded by arable fields, think not of its isolation, but of the scores of thousands of men, women and children who used it in the past. Any isolated church you come across merits attention; it was once an extremely important nodal point.

Very few churches were ever built in isolation. Ecclesiastical authorities needed their local representative, the parish priest, to be at the very heart of community life. The settlement or village grew up around the church; if those dwellings have gone, we can usually find the cause in harsh economic necessity, though in a very few cases the megalomania of a local lord of the manor swept entire communities into oblivion. Historical records will almost always provide the answer. In the case of the church in the photograph, another village, with its own church, and with better links to what became the main highway passing through the district, took root at a place where the old road past the church formed a crossroads with the busier highway. The first settlement withered away over about a century after 1450. Many of its buildings - wattle and daub houses, timber shelters for animals, pens for geese - provided a useful source of construction materials and firewood for the new village, but nobody would damage or rob from a church in those days. Hence, many disused churches have enjoyed a long survival in splendid isolation.

Fig.22

23. Farm Gates

I'm old enough to remember farm gates considerably more elaborate than that shown in Fig.23. They used to have one or two waist-high timber platforms where the farmer placed large metal churns ready for collection by a milk lorry in the early hours of the morning. It's significant that I couldn't find such a gate on the day I went looking for photographic opportunities. Times have changed and milk is no longer collected in that manner.

Another activity that went on at farm gates long before my time has also ceased: the hiring and paying off of day labourers around harvest and crop-picking times. But not to worry; the precise locations of those busy farm gates are clearly recorded on old 6in OS maps. A detector survey of roadside verges, as well as the ground immediately inside the gates, often turns up low denomination Victorian coins and other items. You can locate even those gates that have vanished altogether. Just find the old track that leads directly from the gate to the farmhouse set back off the road. A very straight track often indicates a farm established in Georgian or early Victorian times. If it's a winding track through arable fields, parts of the farmhouse might date from medieval times. The soil behind the hedges could hold hammered coinage because the track might still skirt the area known as the headland where oxen were harnessed and turned when ploughing ancient fields. Plenty of losses occurred at such places.

On narrow public lanes the gated entrance to the main farm in the locality often became a nodal point. When rural bus services began in the 1920s bus stops were located at such gates because there was usually space for the driver to pull-in while passengers clambered aboard or alighted. Horse-drawn charabancs might have done likewise in Victorian times.

Parcel delivery and collection services might also have used the spot in the days when it was safe to leave a parcel in such a public place. Friends, lovers, and children all arranged to meet at locations such as this in a rural world that has now all but vanished. I've noticed that modern farmers struggling to increase their incomes by selling produce at the farm gate frequently select the very gates I

Fig.23

have been discussing here as good spots to set up stalls because the wide entrances allow private car drivers to pull in and make purchases. Casual losses of small change probably still occur, adding to the earlier accumulations around many once-busy farm gates.

24. Monastic Sites

In 1279, the year in which Edward I put a range of new coins, including groats and farthings, into circulation to meet a general rise in the standard of living and a growth in local markets for food products, the Church authorities in England owned one-quarter of all the agricultural land. They had acquired it over many centuries, not by war and conquest, but as a succession of gifts from kings, queens, princes and barons trying, usually on their deathbeds, to bribe their way into heaven. In those days the only way an intelligent young man with no connections to the nobility could get on in the world was by becoming a monk or friar. The life gave him access to written language and to the accumulated works of ancient writers whose books on agriculture, medicine and science were mostly in monastic libraries. So the medieval church had more than vast acres of land; it had all the most knowledgeable farmers and the best-run farms in the country. In southern England, for example, it was commonplace to have vineyards, orchards, vegetable gardens, arable ridge-and-furrow and rich meadows, all on a single monastic farm. Excavation of ancient building remains on several of these have shown that a typical enterprise usually had one or more watermills, one or more stone-built barns, a brick dovecot, a kiln for drying peas and for malting barley, timber henhouses, pigsties and ox-byres, as well as fishponds and fish-traps on the riverside.

Although personally dedicated to vows of poverty, the monks were in daily contact with the cash economy. Each farm employed dozens of lay workers from surrounding villages who received pay for their labour. Surplus produce went to market and the monks came home carrying purses bursting with money. A lot of their profits were sent to the order's headquarter to swell funds for cathedral building projects and other major works. Cash also went to hospitals, and large sums were handed out as charity to the deserving poor who came in a steady stream to the farm gates. Nevertheless, the abbot or friar in charge invariably had a chest brimming with coins at the end of the farming year.

Times changed for the worse in 1348 when the Black Death struck. Half the people of Britain died, and the population figure was not to reach its 1348 level again until 1600. The effect on the farming landscape was obviously dramatic. Demand for corn, oats and barley fell through the floor; huge numbers of plough oxen went for slaughter; farm labourers who survived the plague refused to plough, sow, hoe and harvest for anything less than thrice the regular wage. And, just as today's farming disasters all seem to come at once, the Black Death struck at a time when the climate had begun to deteriorate and soil fertility started to decline after centuries of intensive arable cultivation.

As a result of all those pressures a massive retreat from corn growing occurred. Ploughed fields on monastic farms throughout the land were hurriedly converted to grazing. In Warwickshire, for example, the pasture-to-arable ratio on church lands went up by a half, while most monastic holdings in Dorset and Somerset became sheep-walks, and one-third of Leicestershire properties were soon clothed in grass. What saved ecclesiastical agriculture from utter ruin was a pool of written and practical knowledge. Skilled and literate monks knew all about sheep farming for wool and cattle farming for dairy produce. By 1400 monastic farms were producing fleeces and cheeses by the million. A century and more of accumulating vast wealth was to follow, with money available to tackle major projects such as drainage and woodland clearances that enabled sheep and cattle enterprises to grow ever larger. Paradise on earth seemed to beckon - until Henry VIII grabbed the lot between 1536 and 1539 in what many Roman Catholics still call "The Great Plunder".

Fig.24. shows part of a devastated monastic site that suffered such a fate, it's rooms stripped at the time of any portable treasures, its monks scattered, its fabric robbed in subsequent years of stone, timber, lead and window glass. Fortunately for our purposes there's enough still visible to warrant a mention on a modern Explorer map as Monastic Remains, and the map also shows a web of local footpaths and a stream no doubt harnessed by the monks as a source of power and a source of fresh fish. In other words, there's sufficient information on our maps to guide us to the nodal points around literally hundreds of sites similar to this one

Fig.24

up and down the land. Note that my photograph shows a crop of wheat in what has since become a large field surrounding the visible remains. Clearly the loss of soil fertility in the 13th century was regained in subsequent centuries thanks to the manure supplied by all those sheep and cattle. You'll find that very many former monastic sites now lie in the middle of arable fields. That's good news because even if the structures are now scheduled ancient monuments, you might still get onto arable fields nearby to recover non-stratified losses in the plough soil. In view of all the wealth once attracted to these sites those losses should be numerous.

There's one other point worth making about the photograph. It shows that because stone rubble in the ground made ploughing impossible, nature took over and wild vegetation thrived. Clumps of trees like these in the middle of arable fields should always be investigated. Even if you have no documentary evidence for monastic remains there could be something more interesting that rabbits lurking in the undergrowth.

Northern readers will know of the huge sheep farming enterprises once run by Cistercian monks in Yorkshire, the Lake District and Lowland Scotland. They established many Grange Farms as satellites to their abbeys, each grange tending thousands of sheep. Most suffered the same fates as southern monastic farms during Henry VIII's reign, but unfortunately for our purposes a lot of medieval grange land remains as pasture today and so not so interesting from a detectorist's point of view. The situation may alter if climate changes, or the effects of recent livestock disease, or even the rules of the Common Agricultural Policy, make arable farming more profitable than pasture in future years. If you ever see a plough in those regions follow it!

25. Rural Allotments

Have a look at Fig.25. and then briefly turn to illustration Fig.12b. in the section headed "Converging Routes". That fragment from a Victorian 6in OS map shows very clearly the long, narrow gardens at the rear of a row of village houses. I explained that such gardens were once called tofts - patches of ground allotted to medieval villagers for growing their own food and erecting timber sheds where their oxen slept. I might have added that in many villages where there is not too much pressure to find new house-building land, the tofts still exist as allotments, most now tended by village horticultural enthusiasts.

What I find interesting about Fig.25. is, firstly, that the modern villagers seem to have kept to the medieval strip layout, as shown by the lines of bamboo canes. Even more interestingly the strip of ground closest to where I stood when taking the photograph matches the medieval headland where a plough-team of oxen were turned before setting off to plough another furrow. I suspect that losses covering many centuries will lie beneath the headland's grass, though the toft strips, which have probably been hand cultivated for just as long, must also hold coins and relics reflecting village history.

If you come across allotments similar to this check on your Explorer to see if another footpath or cart track continues the line of the headland and runs out of the village and into nearby arable fields. This might be the very route taken by medieval ox-teams and men on their way to work each day. Even though the ancient headland will have been ploughed up, the true line of any footpath across modern arable will indicate its position. Concentrate on that part of the modern field if you obtain search permission. It's where men and their animals would have rested during the day centuries ago.

Fig.25

Never neglect an opportunity to win detecting permission on rural allotment sites such as this. Apart from the strong possibility of good finds, you will also meet many villagers who know a lot about local fields. Someone might be prompted by the sight of your detector to recount a story about "old coins" ploughed up by his grandfather. He might even remember the exact location. Any allotments located within a few yards of older village houses, whether or not they now have the sort of layout just described, deserve your consideration. They are certainly nodal points now; they may have been for generations.

26. Rivers & Streams

We owe a great debt to Raymond Selkirk, author of **The Piercebridge Formula** and **On The Trail Of The Legions**, for opening our eyes to the importance of rivers as part of the ancient communications network of Britain and the rest of the Roman Empire. His researches and subsequent books showed clearly that stretches of water we now regard as little more than damp patches trickling across the countryside were once strongly flowing rivers perfectly capable of carrying small commercial vessels loaded with a ton or more of cargo and a crew of perhaps two men and a boy. Deliveries of such heavy loads via long-distance roads would have proved impossible 2,000 years ago, even if bullock carts had been allowed to use Roman military routes. My own researches while writing **Reading Rivers** showed that as recently as a century ago numerous tributary streams were still capable of carrying commercial loads (see page 19 in that book). Railways and then lorries, finally killed them off, but evidence and artefacts from up to two millennia of human activity on and alongside numerous rivers and streams lie there to this day.

One of the first research tasks to undertake when you spread out an Explorer map is to locate all rivers and streams, marking every place where a road, bridleway or footpath touches the banks. A day spent walking or cycling parallel with a river or stream that has a dozen or more of these nodal points will provide you with some interesting sites worthy of further investigation. The intersections will include:-

Bridges - If there's no date-stone on a bridge, judge its age by getting underneath the arch if possible. Down at the water's edge you will have a revealing view of original stonework that, on any bridge more than 150 years old, will be large compared to stones used in modern widening work or in raising the height of parapets. When you feel sure the bridge is at least 200 years old look for nearby flat fields on both sides of the river. If the fields have old gateways aligned with the bridge it's very likely the fields had some sort of commercial connection with the bridge. Detector searches might tell you more. Check local footpath alignments to find a field where lots of people congregated near the bridge.

Fords - Two or more aligned footpaths approaching river or stream banks on both sides of the water strongly suggest a fording site or stepping-stones across shallow water. Gravel bottoms were preferred to mud or sand when selecting a crossing place. Flanking pastures would once have attracted drovers who liked to meet at fords because a ford usually had gently sloping approaches to the water's edge where livestock could drink. Search nearby fields for drover's losses. Wait for drought conditions, and then detect the gravel riverbed.

Ferries - If a deep river crossed an important ancient or medieval route at a location some distance from a major settlement the most common solution to the problem was to introduce a ferry service. Bridge building required large sums of money; a ferry usually required no more than one (but sometimes two) small boats and a rope or chain stretched from one bank to another, plus a couple of ferrymen. On pilgrims' routes monks often ran ferry services and took the opportunity to sell pilgrims' badges to their passengers. Check all available fields adjacent to ferry sites confirmed by historical research.

Sheep washes - In former (or present-day) sheep rearing country look for roads, footpaths and sheep walks that converge on a stream or river bank where the water was/is fast-flowing. Such places were used communally by shepherds and their families to give the flocks a pre-shearing bath that would add value to each animal's fleece. The flocks were usually driven into a deep pool; the shepherds followed them, making sure that every animal was totally submerged and scrubbed. Everybody in the locality turned out to share the fun and the occasion, which was a good excuse for drinking ale and making merry. Grassy banks flanking the river would be packed with people and a nearby field would probably be used as an impromptu market site where sheep changed hands, wives sold homemade food, and daughters looked for prospective husbands.

Fisheries - Salmon, trout, eel, perch, chub, gudgeon, lamprey, grayling, charr, and crayfish were all part of a thriving freshwater fishing industry up to early Victorian times when industrial and agricultural pollution began to take their toll. In pre-Victorian days almost all rivers and streams in Britain had good stocks of one or two species, while larger rivers teemed with every species. Catching them called for the deployment of fish weirs, which were multiple lines of tightly packed stakes driven into the river or streambed with the aim of entrapping any fish that tried to swim through the barrier. They caught fish very efficiently, but they also interfered with the free passage of riverboats, and sometimes with the flow of water to millraces. Friction between boatmen, millers and fishermen was exceedingly common in medieval Britain. Records of cases that ended up in court can be useful because they often record the precise location of the offending weir. The banks alongside it were probably the site of a fresh fish market; if the ground remains accessible today, it should hold plenty of lost coins. Fishing with nets was also practiced, especially on rivers where salmon and eel were plentiful. Again, look for bank-side fish markets.

Farmed fish kept in man-made riverside ponds, or in ponds made by partially blocking streams, ensured healthy fish and steady supplies of Friday eating for monks. Large manor houses also had their own fish ponds sited alongside rivers or streams. Stonework from these survives long after the ponds have dried up. Detecting around the stonework should yield interesting finds.

Locks - In medieval times a river lock was often no more than a gap in a fish weir, or a sliding gate in a timber dam. When opened, it provided a surge, or flash flood, of water to get vessels over shallows in the river, hence the common name "flash lock". Seek flash locks near water mills because riverbed conditions ideal for millraces were also ideal for flash locks. It is also worth searching flanking banks, likely to have been busy places in the past.

Fig.26

Sharp bends - At a sharp bend in a river the banks often widen, giving more room for vessels to turn around. This makes sharp bends good stopping/loading/unloading points. If footpaths and tracks converge at a bend, check flanking fields for losses.

Springs - There's a spring at the source of every river and stream. In ancient times some became sacred sites, attracting devotees and their offerings. Nearby footpaths and fields should be carefully searched.

Stream sides and river banks - Sometimes footpaths converge at a place flanking flowing water simply because the spot offers pleasant views. More often the convergence tells us that the place was busy in the past with vessels loading, unloading or launching. The ground around such locations could be rich in losses.

Watermills - Fig.26. shows a watermill on a once-busy river. The mill, now converted to a domestic dwelling, has a large rear garden with two footpaths converging on a back-gate. This was probably where wagons and/or mules waited with their sacks of corn, and then loaded flour. Money probably changed hands there. A search might bring losses to light.

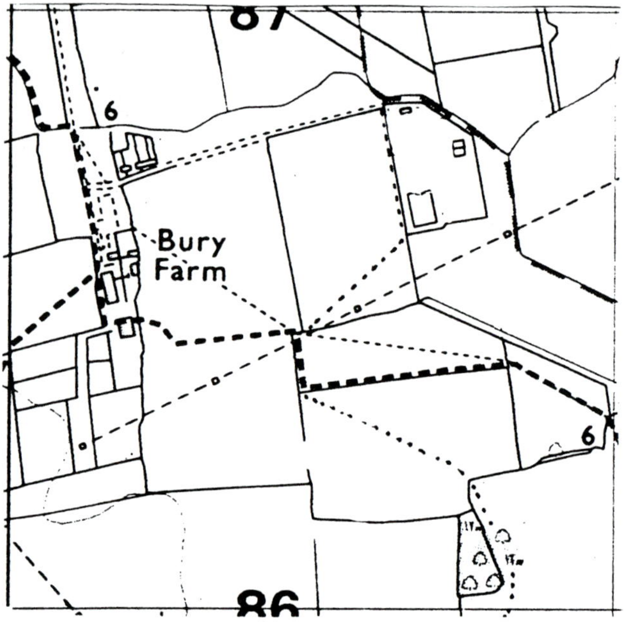

27. Named Meeting Places

Look at the snippet from an old 6in map reproduced as Fig.27. It shows half a dozen footpaths converging on....well, what looks like no more than the point at which three fields meet at a corner, presumably where a stile provides the means of getting over a fence. But look again at the name of the farm just one field away: Bury Farm. That's a clear indication that at some time in the past there was a prominent ancient burial mound in the vicinity. The bury, as they were called in some parts of the country, long ago fell victim to the seasonal attrition of a ploughshare, but folk memory kept the name alive in a local farm, probably because the bury received mention in ancient legal documents about land ownership.

Tens of thousands of burys, barrows, howes, lowes, motes - to use some of the better-known dialect words for these sites - have been identified across Britain, some as existing mounds, most as no more than crop marks. The vast majority were robbed of any grave goods by Victorian antiquarians, but when the mounds were still visible they were used as meeting places, beacons, bases for Christian crosses, windmill platforms, even as gibbet sites. Crowds converged on them in the past. Anglo-Saxon lawyers used them as reference points when setting the boundaries on parcels of land. Medieval farm labourers resorted to them for courting and other village games. Georgian militia regiments used them as targets for musketry practice. Victorian politicians and evangelists used them as outdoor podiums from which to preach or make rash promises. Old footpaths converging on them confirm not only the location of the ancient mound, but also the importance of the location as a nodal point. Soil for several hundred yards around a bury should be thoroughly searched.

Other farm, field and place names hint at locations where people gathered in the past. If any of those converging footpath locations you marked on your Explorer maps lie close to places with one of the following linguistic elements in their names, your footpath clues become even more likely to point to a spot where numerous coins and relics were lost in the past. Look out for: -ale, -ansty, -bank, -char, -cote, -cross, -fair, -ford, -gospel, -halt, may, -merry, -moot, -money, -oak, -penny, -port, -rest, -smith, -stone, -stoup, -thorn, -tun.

Fig.28(a)

28. Farm Shops & Pick-Your-Own Sites

Never dismiss present-day locations with Pick-Your-Own signs at their gates as "too modern" just because your aspirations lie in picking ancient and medieval coins and relics out of the ground. Any invitation to buy farm-produced vegetables or meat offers a golden opportunity to introduce yourself to the farmer and to do your prospects of getting onto fields holding ancient losses a power of good. At PYO fields buy some of whatever crop is seasonally on offer, then, when you line up to pay, mention to the farmer that you're a detectorist and you'd be more than happy to come to the aid of any customer who reports the loss of a ring while picking strawberries or peas. Believe me, it's a very common occurrence. It happens so often that the farmer probably already has a list of customers' names and phone numbers with a note of where each lost some valued possession. Tell the farmer that the best time to search the

Fig.28(b)

picking field will be when he ploughs and harrows it; and don't forget to ask if you can come to some arrangement with him about any older losses that come to light. Word soon gets around in rural communities, but if another local farmer doesn't say that he's heard of your good services at the PYO farm, make sure you tell him while seeking permission to search his fields.

Intensive crop cultivation requires good soil, aspect, drainage and fertility, so it's very likely that PYO fields lie in some of the parish's prime agricultural land. If ancient finds have come to light in the past, how do the locations tie in with the fields you have gained permission to search for modern losses? Integrate all the information you now possess about local footpaths and you should have excellent prospects ahead. Even if there are no written records of ancient discoveries, your new friends in the local farming community may have valuable information to pass on to you.

Fig.29

29. Motte & Bailey Sites

You may not come upon a Norman motte as prominent and so well cared for as that shown in Fig.29., but the known locations of numerous long-vanished motte and bailey sites are well documented, with one or two probably marked on your own Explorer maps. If the site of one now looks like no more than a slight bump on the ground surface, let me remind you that the original bailey comprised a courtyard with a hall, various domestic buildings, stables, workshops and a surrounding ditch, with earth thrown up to form a wall on the inside, atop of which sat a strong timber palisade. In peaceful times, the lord, his knights and retainers, together with servants and trusted locals, all lived within the bailey. The motte, to which everyone retired when danger threatened, was a much higher mound alongside the bailey. It had its own ditch and an even higher mound with a timber watchtower above. The last vestiges of this highest mound remain visible at many sites today. Motte and baileys date from the early Norman period, but because selection of the location was determined by defensive considerations, many were abandoned in favour of better agricultural prospects within a few years. Nevertheless, they acted as important nodal points in the late 11th century, and local fields adjacent to the former bailey will probably hold early Norman and perhaps late Anglo-Saxon losses.

Fig.30

30. Modern Farming Techniques

Just before the main harvesting season you'll spot an array of agricultural equipment in most farmyards. The farmer and his men might be tinkering with engines and machinery, so this can be an appropriate moment to try a direct request for permission to search the bare soil on each field as soon as crops are harvested and a tractor and rig have turned, tined and harrowed the ground in preparation for the next seeds. When the farmer knows you a little better he might answer questions about how deeply he intends to plough his various fields. The information is crucially important to you because it is the ploughshare more than anything else that will bring ancient and medieval losses within range of your detector. You can never know enough about what the machinery on the farm will be doing to the soil during the next 12 months, but you must strike a careful balance between information gathering and pestering.

31. Vanished Mansions

Exploring along a country lane one day you might suddenly encounter an incongruously large entrance to what might be an imposing building....only to find little more that bare fields or a very modest modern house when you stop to look through the gates. You've probably stumbled upon the site of a vanished mansion, perhaps dating from Georgian or early Victorian times. Profits from the Industrial Revolution gave mill and factory owners in grimy towns the riches to satisfy their heart's desires. Many opted for playing the role of country gentlemen. They bought land and spent lavishly on building houses that required small armies of servants to run them, backed by gardeners, farm labourers, grooms and other staff to supply food to the kitchens. Some of these industrialists managed to integrate into county society, but many more failed to get onto the invitation lists of neighbourhood blue bloods who could trace their occupation of the land back to William the Conqueror. In response to the cold shoulder some newcomers spent even greater sums on their country houses, including the erection of large gateways, as in Fig.31. The Depression years following the First World War sent many owners into bankruptcy. The servants went to cities and towns; the houses fell into decay; some were pulled down in later years, often leaving one or two outhouses with ornate brickwork, or (as here) a wide entrance lacking its gates. Introduce yourself to the present landowners and offer to search what were the kitchen gardens and former lawns of the old house. You might make some valuable finds

Fig.31

32. Prosperous Villages

As we have seen so dramatically in recent years, agriculture has its economic downs as well as occasional ups. The wheat barons of the 1960s-70s made fortunes from surplus mountains of grain and spent their wealth buying properties and farms in warmer climes. Today's farmers, some scraping by on state handouts, are tightening their belts to the last notch, perhaps hoping that a detectorist will find a hoard on their fields and turn fortune around. Back in the 15th century, in the decades following the Black Death, a steady recovery from deep depression was taking place. Corn prices rose every year, but many farmers were more than happy to stick with sheep rearing because the boom in demand for wool was pouring ever larger sums into rural economies across England. As a result, yeomen in villages that previously had only one or two fine houses (the squire's and the priest's) began to display their wealth by building new homes. Many such properties survive to this day (see Fig.32.) because they were erected within villages, unlike the isolated Georgian and Victorian mansions thrown up by newcomers to the countryside in later centuries. If you spot architecture like this in a village street try to win detecting permissions on all available land close to the village. Make special efforts where fields that fed sheep until a couple of years ago have recently become arable land. You could find large numbers of medieval hammered, especially if you apply your knowledge of footpath clues to locate the important nodal points.

Fig.32

DETECTOR FINDS 5
Gordon Bailey

Advanced Detecting
Norfolk Wolf
How to improve your techniques & finds rate!

Saxon & Viking Artefacts
Nigel Mills

DETECTOR FINDS 4
Finds Identified
Gordon Bailey

Successful Detecting Sites
Locate 1000's of superb sites and make more finds!

Medieval Artefacts
CATALOGUE AND PRICE GUIDE
Nigel Mills

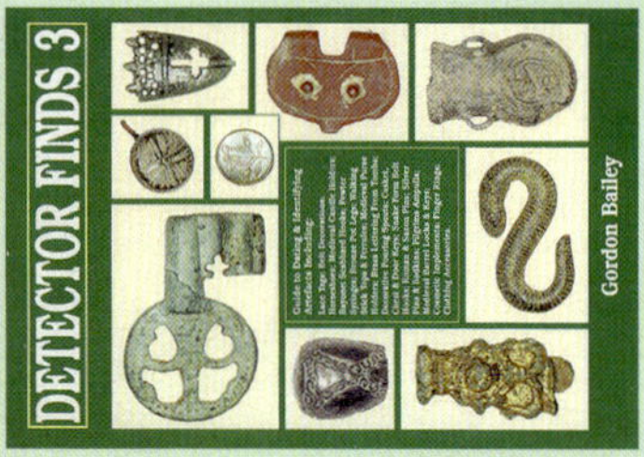
DETECTOR FINDS 3
Gordon Bailey

by David Villanueva
Site Research
for Detectorists, Fieldwalkers & Archaeologists

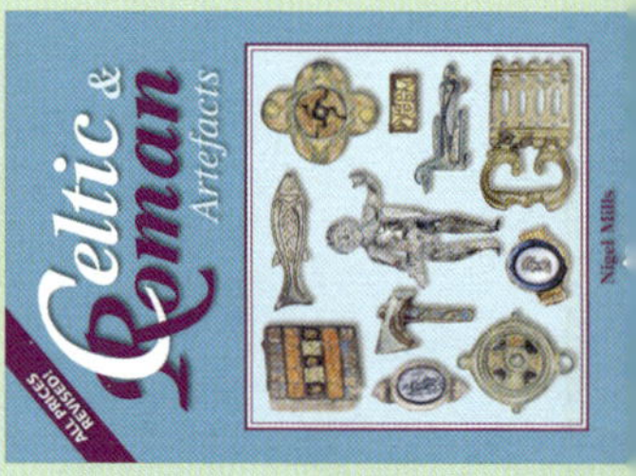
ALL PRICES REVISED!
Celtic & Roman Artefacts
Nigel Mills

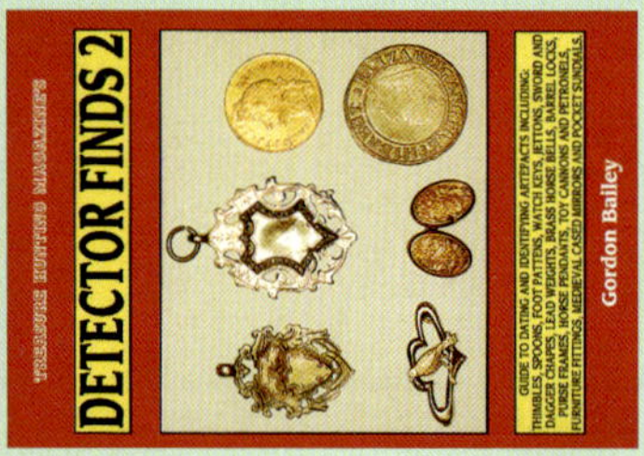
TREASURE HUNTING MAGAZINE'S
DETECTOR FINDS 2
Gordon Bailey

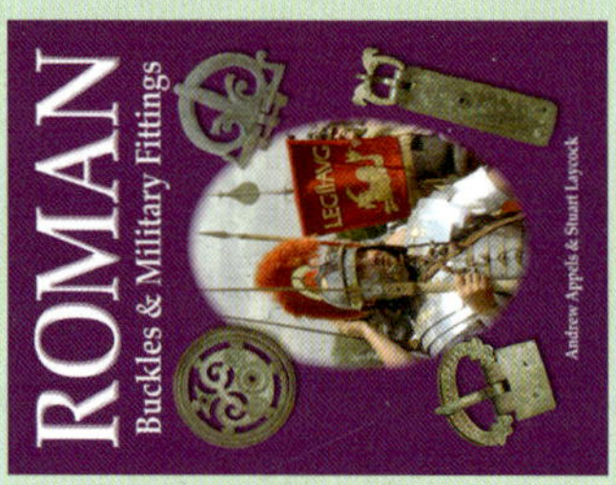
ROMAN
Buckles & Military Fittings
Andrew Appels & Stuart Laycock

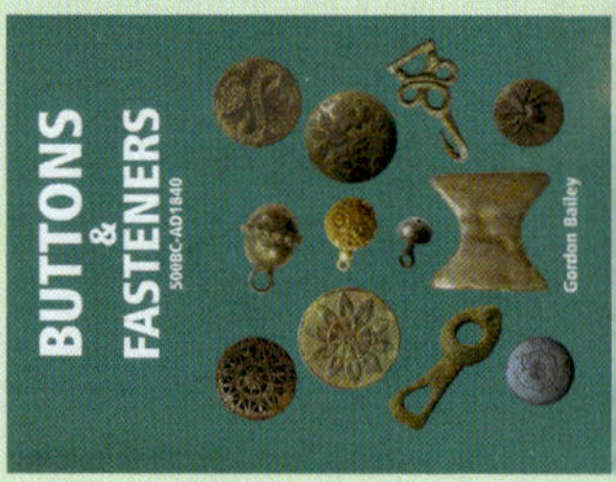
BUTTONS & FASTENERS
500BC-AD1840
Gordon Bailey

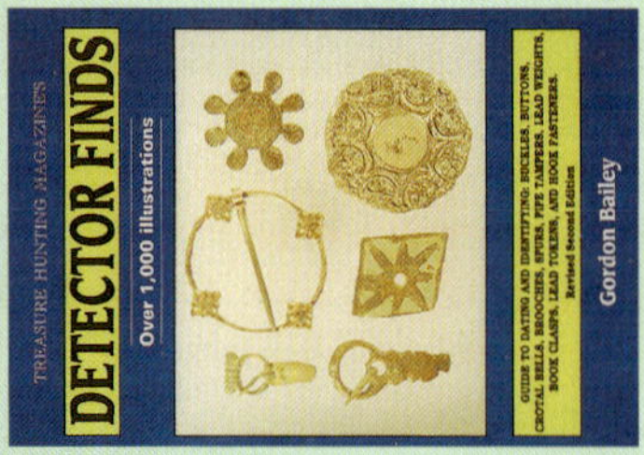
TREASURE HUNTING MAGAZINE'S
DETECTOR FINDS
Over 1,000 illustrations
Gordon Bailey

DETECTOR FINDS 6
Gordon Bailey

BEGINNER'S GUIDE to METAL DETECTING
Julian Evan-Hart & Dave Stuckey